Russian Contributions to Analytical Chemistry

Yury A. Zolotov

Russian Contributions to Analytical Chemistry

Yury A. Zolotov
Department of Chemistry
Lomonosov Moscow State University
Moscow, Russia

ISBN 978-3-030-07532-3   ISBN 978-3-319-98791-0 (eBook)
https://doi.org/10.1007/978-3-319-98791-0

Softcover re-print of the Hardcover 1st edition 2018

This Springer imprint is published by the registered company Springer Nature Switzerland AG
The registered company address is: Gewerbestrasse 11, 6330 Cham, Switzerland

Preface

True science is international, with its achievements representing universal value and not observing national borders. Of course, this primarily pertains to basic science, but also, to a significant degree, applied science, at least as long as military or commercial secrets do not preempt it. However, any country is proud of its scientific achievements and reveres its prominent scientists, especially pioneers.

At the same time, it is known that recognizing somebody's priority in science can cause certain problems. To be fair, there are also pioneers who are happy just knowing they were the first and are not very interested in garnering recognition. However, there are barely a handful of such researchers; most of them trying to demonstrate and prove the novelty of their work, thus making the issue of priority rather significant. Objective judgements about who was first can be made by historians of science, but they are hardly able to deal with all important innovations. Besides, there are usually only a few specialists in the history of a particular science, with the number decreasing even further if one allows for the fact that they should also have expertise in the subject matter, in the science itself. The origin of methods, generalizations, ideas, and so on, is best of all known to the most advanced and, as a rule, seasoned researchers who have been working in a particularly narrow scope. However, this knowledge, on the one hand, concerns not very ancient achievements; on the other hand, not all active scientists are interested in history—some are more agitated by problems on the frontline of science and are more worried about their own research, with its current, all-engulfing challenges. They have things other than history on their minds; history is not their first consideration.

However, one should, after all, pay tribute to those who deserve it. Besides, securing the recognition of discoveries is of great value for the team, the region (organization, city, country) where the pioneer works or used to work, and the distributed scientific community that the scientist belongs or used to belong to. Natural pride, a sense of belonging to a common effort, and craving to be successful, well up in the members of such a team, community, etc., with all of them being very powerful incentives. Therefore, this needs to be dealt with so that scientists are treated according to their merits.

In this book, such analysis and assessment is attempted in regard to those who developed and are developing analytical chemistry in Russia. The book consists of ten chapters and tries to encompass all major methods and trends.

I will be grateful to anyone who reports omissions and inaccuracies in the book and lets me know if any of my judgments were not quite correct, since after all I am no specialist in all the aforementioned methods and trends. I would like to express my sincere gratitude to those of my colleagues who took time to read, comment on, and critique chapters of interest to them, namely, corresponding member of the Russian Academy of Sciencies, E. N. Nikolaev and doctors of sciences, M. A. Bolshov, G. K. Budnikov, M. V. Gorshkov, A. T. Lebedev, M. A. Proskurnin, I. A. Revel'sky, G. I. Tsyzin, and candidate of sciences A. I. Kamenev. Special thanks go to Olga Igorevna Popova and Natal'ya Vladimirovna Gracheva for their help in prepress preparation of the manuscript.

Moscow, Russia Yury A. Zolotov

Contents

Chapter 1
Spectroscopic Methods

Abstract Russian scientists developed electrothermal atomic absorption spectrometry (B. V. L'vov), made a significant contribution to luminescence analysis (the notion of a quantum yield, the Shpol'skii effect, etc.), proposed X-ray polycapillary optics (M. A. Kumakhov), developed the X-ray radiometric method (A. L. Yakubovich), and discovered electron spin resonance (E. K. Zavoiskii).

1.1 General Remarks

Spectroscopic analytical methods include methods based on the use of radiation from the entire electromagnetic spectrum—from gamma rays to radio waves. All these methods were, and still are, being developed to some extent in Russia.

In atomic emission analysis scientific improvement has come in the form of the two-jet arc plasmatron used as a source of spectral excitation. Its use has not been particularly widespread, but it has a number of significant advantages for the analysis of solid free-flowing samples. One can also mention the so-called spilling method for the semiquantitative analysis of powders. In the field of atomic absorption spectrometry—this is the creation of an electrothermal version of the method by B. V. L'vov—Russian achievements are well known, and the priority of L'vov is not disputed.

In optical molecular absorption analysis, one should note the contribution to photometric elemental analysis; this contribution consists of the creation, study, and wide use of organic analytical reagents, mainly complex-forming ones, for photometry. One of the best-known reagents is Arsenazo III (S. B. Savvin). Many Russian analysts worked in this field in 1950–1980.

In the field of luminescence analysis, a significant contribution was made by S. I. Vavilov (the concept of quantum yield, analytical methods, etc.), E. V. Shpol'skii (the famous Shpol'skii effect), E. A. Bozhevol'nov, and others. In X-ray methods, M. A. Kumakhov's X-ray optics (Kumakhov lens) is now used in a number of instruments.

These and some other achievements are discussed in more detail below.

Y. A. Zolotov, *Russian Contributions to Analytical Chemistry*,
https://doi.org/10.1007/978-3-319-98791-0_1

In a smaller scale, Russian analysts have studied, and are studying today, vibrational spectroscopy and its use in chemical analysis. An exception, I suppose, is the use of infrared spectrometry in the near field. This method is used for the analysis of agricultural products, pharmaceuticals, and some other objects. Corresponding devices are also produced, especially by "Lumex Co." in St. Petersburg [near infrared field (NIR)]. Work on radio spectroscopy [nuclear magnetic resonance (NMR), electron spin resonance (ESR), and nuclear quadrupole resonance (NQR)] is usually carried out not by analysts but by other specialysts, e.g., organic chemists or physical chemists.

During the 1940s–1970s, in the Academy of Sciences of the USSR, was a Commission on Spectroscopy, which did much for the development of atomic emission and other methods, in terms of their popularization and practical uses. The Commission was headed by G. S. Landsberg, S. L. Mandel'shtam, and other well-known scientists; the Commission even had its own laboratory, on the basis of which were established the Institute of Spectroscopy of the USSR Academy of Sciences and several scientific councils. At present, the Commission on Spectroscopical Methods of the Scientific Council of the Russian Academy of Sciences on Analytical Chemistry regularly convenes conferences devoted to these methods of analysis.

It should also be remembered that electron spin resonance (ESR) was discovered by E. K. Zavoiskii in Kazan in 1944, and Raman spectroscopy was created simultaneously by (1928) K. Raman, on the one hand, and G. S. Landsberg, and L. I. Mandel'shtam, on the another.

1.2 Electrothermal Atomic Absorption Spectrometry

In 1974 Boris Vladimirovich L'vov was awarded the Gold Medal by the Talanta journal; for the presentation of the medal the editor of the journal, M. Williams, came to the USSR. This was the first award B. V. L'vov, received for his creation of electrothermal atomic absorption spectrometry (ETAAS)—other awards followed. He also received invitations to conferences and universities as a lecturer, etc. The importance of B. V. L'vov in the development of the atomic absorption method using a graphite atomizer is generally well recognized. He went on to write about the history of the creation of electrothermal atomic absorption spectrometry [1].

"Having entered the position of junior researcher at the State Institute of Applied Chemistry (SIAC) in Leningrad in early 1955, I was for many years involved in the spectral analysis of materials labeled with radioactive and stable isotopes. Working through the writing of abstracts for the Russian reference journal Khimiya (Chemistry), at the end of 1955 I accidentally ran into Walsh's paper [2]. The possibility of developing a method of absolute analysis, free from the need to use standard samples of the composition, which Walsh mentioned, seemed so attractive to me that I decided to devote all the free time from basic tasks to this problem. In the summer of 1956, taking advantage of my vacation and the absence of other employees in the laboratory, I conducted the first experiments on the visual observation of the absorption of sodium D-lines. For this purpose, I used a demountable tube with a hollow cathode, which I prepared for carrying out isotopic analysis, a graphite tube furnace heated on a stand for fractional distillation of volatile impurities from non-volatile matrix, and a prism monochromator. The impression of gradual fading and the complete disappearance of bright sodium lines as the furnace was heated appeared to be so stunning that it determined my scientific interests and to a large extent my personal life for many years.

The reaction of the staff of the spectral laboratory was far from inspiring. Even G. I. Kibisov, the well-known spectroanalyst in the country, who supervised my activities at the institute, repeatedly expressed concern about my risky interest.

Nevertheless, during the year I managed to prepare a very cumbersome laboratory installation for carrying out quantitative measurements. The choice of an isothermal furnace (graphite cell) as an atomizer, as one can see from the notes in the working journal (Fig. 1.1), was sufficiently conscious. It was not difficult to figure out that AA measurement of the absolute magnitude of the analytical signal is only the first condition of the absolute analysis. Another problem was the creation of a suitable atomization technique capable of guaranteeing full evaporation of the sample, atomization and retention of vapor in a certain volume inside the furnace (like a cell in absorption molecular spectrophotometry).

Using this setup (Fig. 1.2), it was possible to demonstrate the undoubted advantages of the graphite cell before the flame in the sensitivity, the coverage of determining elements (the C_2H_2,/N_2O flame was not yet known) and even in the possibility of the absolute analysis. These results were reported at the All-Union Congress on Spectroscopy in October 1958 in Moscow and published in 1959 in the Engineering Physics Journal [3, 4]. However, there was no appreciable interest in this method among spectroanalysts. According to the Institute of Scientific Information (ISI, Philadelphia), the papers [3, 4] were quoted only three times for 6 years (1960–1965). In part, this was due to the fact that the journal was not well known among spectroanalysts. A little more (11 times in 4 years) was cited an article [5] published in 1961 in Spectrochimica Acta. Nevertheless, even leading spectroscopists of the country (Prokof'ev, Zaidel', Mandel'shtam, Nedler, etc.) were sufficiently restrained (rather critical) in assessing the prospects of the method. Only six years after the start of the research, I managed to defend my Candidate of Sciences thesis not without difficulty.

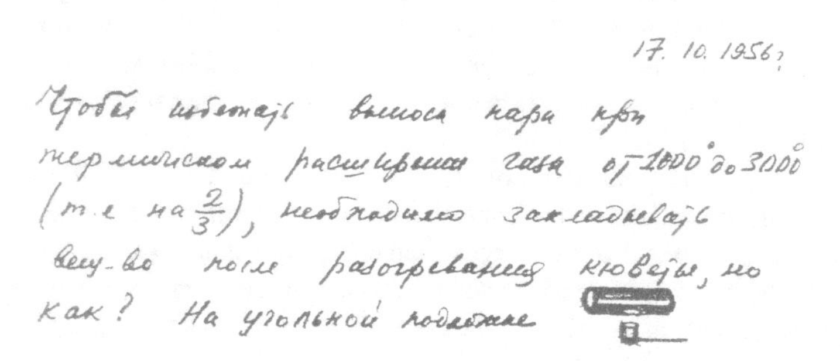
17. 10. 1956 г.
Чтобы избежать выноса пара при термическом расширении газа от 1000° до 3000° (т.е. на 2/3), необходимо закладывать в-во после разогревания кюветы, но как? На угольной подложке

Fig. 1.1 A record in B. V. L'vov's workbook made on October 17, 1956 regarding the first experiments on electrothermal atomic absorption spectrometry. With permission of the Russian Academy of Sciences

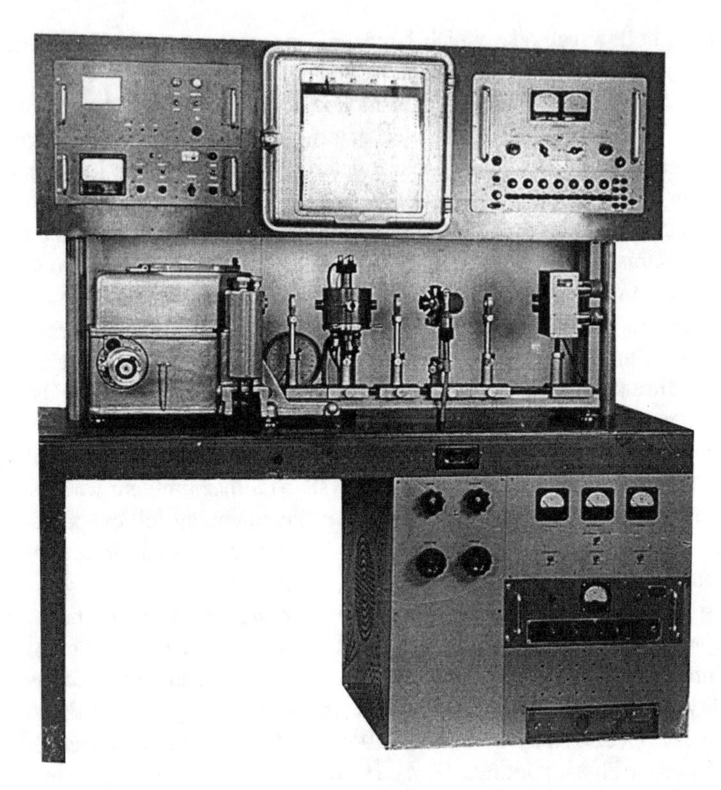

Fig. 1.2 One of the first of B. V. L'vov's devices realizing an electrothermal version of atomic absorption spectrometry. With permission of the Russian Academy of Sciences

The research was hampered not only by the lack of moral support from the scientific community, but also by the fact that for 10 years I had to work alone, without the help of other employees (the "initiative" topic did not fit into the official theme of the institute). I had to do everything myself—from designing chambers for the use of graphite cells to grinding consumable graphite parts (furnaces, contacts, and electrodes). In addition, the SIAC dealing with defense research on rocket fuels was completely closed to outsiders (especially foreigners). Employees of the institute, under the threat of dismissal or more severe sanctions, did not have the right to go abroad, meet or correspond with foreigners, publish abroad without the permission of the ministry, etc. This also did not facilitate the exchange of ideas and the improvement of the method.

Despite this, during ten years of efforts I managed to significantly improve the technique and procedure of the analysis. For the first time in AAS electrodeless discharge lamps were used for a large group of highly volatile elements, the coating of furnaces with pyrolysis graphite and the increased argon pressure in the atomizer were proposed and used. The arc heating of the electrode with the sample was replaced by a simple ohmic method, and a method for measuring the integral absorption (the area of the impulse) was proposed and justified. For the first time in

ET AAS an automatic metering scheme of non-selective spectral interference with the help of a deuterium lamp was used. These improvements were described in the series of articles [5] and in the monograph [6], revised and translated into English [7], and also embodied in the design of a laboratory AA spectrophotometer with graphite cell.

An important milestone in the further development of the ET AAS was the publication in 1966 of an article by Hans Massmann [8], who worked at the Institute of Spectrochemistry and Applied Spectroscopy in Dortmund. The circumstances that contributed to the emergence of his interest in the ET atomizer (ETA) are curious. After participating in the Second World War, he spent several post-war years in the prisoner of war camp near Moscow, where he learned Russian. On his return to Germany, this made him possible to get acquainted with the author's first Russian publications earlier than anyone else abroad [4, 5]. Being a more pragmatic than his predecessor, Massmann significantly simplified the furnace design and the analysis procedure by replacing the evaporation of the sample into the isothermal furnace with a help of an additional electrode for the evaporation of the sample from the furnace wall during the heating thereof …"

Boris Vladimirovich L'vov *was born on July 9, 1931. He graduated from Leningrad State University (1955). He is a doctor of physical and mathematical sciences, a professor, and was head of the Division of Analytical Chemistry of Peter the Great St. Petersburg State Polytechnic University. He is an doctor honoris causa of Strathclyde University (Glasgow, Great Britain), a member of the editorial boards of a number of international journals, and worked in a number of International Union of Pure and Applied Chemistry commissions. Professor L'vov was an honorary member of the Scientific Council of the Russian Academy of Sciences on Analytical Chemistry (SCAC). He was awarded: the Gold Medal from the journal Talanta (1974), the medal of Cardinal Leme of the Catholic University of Rio de Janeiro (1988), the James Waters Prize, the Bunsen–Kirchhoff Award from the German Chemical Society, the Marcus Marci Medal from the Czech Spectroscopic Society, the Gold Medal of the 30th International Colloquium on Spectroscopy, and the SCAC Award* (Fig. 1.3).

His area of scientific interest was atomic absorption spectroscopy. He invented ETAAS. He proposed the idea of atomization of substances based on the complete evaporation of a sample in a miniature graphite furnace, in addition to many other technical and methodological improvements, which ensured high sensitivity and precision of analysis. He developed the theory of ETAAS and suggested ways of applications of this method in fundamental research. He also proposed and justified new approach to the interpretation of kinetics and the mechanism of decomposition reactions, based on gasification of compounds with simultaneous condensation of non-volatile decomposition products. He also proposed a gas-carbide mechanism for the reduction of oxides using carbon. He authored more than 300 scientific works, including 2 books.

In the first serial devices that realized the electrothermal version of atomic absorption spectrometry, produced by Perkin Elmer, a Massmann furnace was used. However, the matrix effects associated with the use of this furnace were too great,

Fig. 1.3 Boris Vladimirovich L'vov (born July 9, 1931), the creator of electrothermal atomic absorption spectrometry. For many years he was in charge of the Department of Analytical Chemistry of Peter the Great St. Petersburg Polytechnic University. The most quoted Russian analyst. Photo provided by Prof. B. V. L'vov

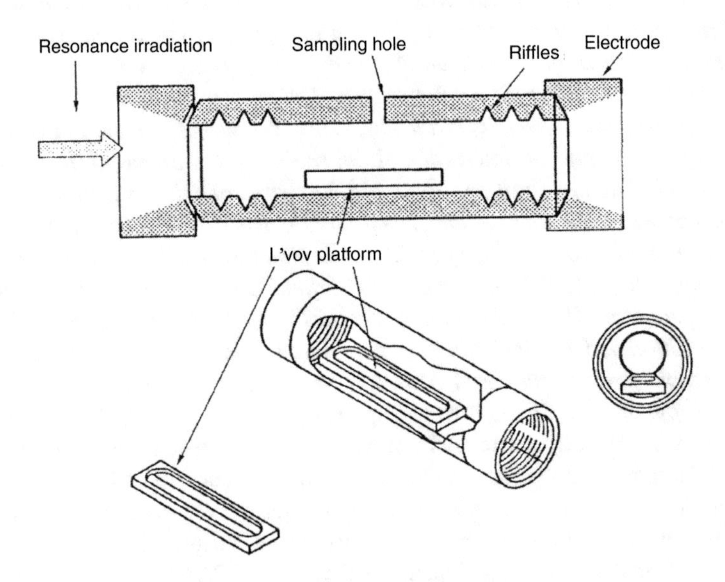

Fig. 1.4 Massmann furnace with L'vov platform. With permission of the Russian Academy of Sciences

and after a while B. V. L'vov carried out some improvements—he introduced a small platform into the electrothermal atomizer (Fig. 1.4), which improved conditions for analysis, accuracy, and sensitivity [9]. The platform was used very widely and was known as the "L'vov platform."

1.3 Fluorescence Analysis

The fundamentals of luminescence as a phenomenon were laid down by J. G. Stokes (1819–1903). For chemical analysis, luminescence has been actively used since the mid 20th century. In the development and application of fluorescence analysis much was done by physicists and chemists who worked in the USSR.

An important contribution to the physical foundations of fluorescence analysis was made by S. I. Vavilov. He measured the fluorescence yield as a function of exciting light wavelength, investigated how the ratio of the fluorescent radiation energy to absorbed light energy is changed when fluorescence is excited by light of different wavelengths. It turns out that energy yield increases linearly with an increase in wavelength of exciting light. In other words, the fraction of absorption of light energy given in the form of fluorescent radiation increases as the wavelength of the exciting light approaches the wavelength of the fluorescent radiation. This means that whether fluorescence is excited by shorter wavelength light (e.g., ultraviolet) or longer wavelengths, with quanta of lower energy, the same number of quanta of fluorescent radiation always accounts for the same number of absorbed quanta; consequently, the quantum yield is constant. Therefore, the smaller the difference between the emitted quantum and absorbed one (that is, the closer the wavelength of the exciting light to the wavelength of the fluorescent radiation), the closer to unity the ratio of the energy of the emitted radiation to the absorbed radiation. The found constancy of the quantum yield was not a random empirical regularity. Vavilov identified, within this relationship, the fundamental law of photoluminescence, and thereby revealed the essence of the phenomenon. The very concept of quantum yield was introduced by Vavilov [10].

Later S. I. Vavilov and his school also found other regularities in the conversion of absorbed light energy into fluorescent radiation, which created more fundamental prerequisites for the development of fluorescence analysis [11], e.g., resolving the problem of determining ozone in high layers of the atmosphere, prior to work undertaken by geophysicists, as well as estimating the distribution of ozone at various heights. S. I. Vavilov outlined a variety of ways to use fluorescence analysis. This was in the 1930s, when the very possibility of using fluorescence analysis as a method of chemical analysis was considered controversial. He also paid much attention to organizational work on the development of fluorescence analysis (hardware provision, publication of manuals and abstract collections on fluorescence analysis, and convocation of meetings).

Sergei Ivanovich Vavilov *(1891–1951) was an optical physicist. He was an academician (1932), a doctor of physics and mathematics, a professor, and four times laureate of the State Prize of the USSR. He graduated from Moscow University in 1914. From 1914 to 1918 he completed military service. He then worked at the Moscow University, Moscow Higher Technical School, the Institute of Physics and Biophysics, the State Optical Institute, and the Physical Institute of the USSR Academy of Sciences. He was president of the USSR Academy of Sciences between 1945 and 1951, and participated in the creation of the Moscow Institute of*

Fig. 1.5 Sergei Ivanovich Vavilov (March 12(24), 1891–January 25, 1951), an academician, an optical physicist, who made great contributions to luminescence analysis. https://www.lebedev.ru/ru/personalities/u-istokov/452

Physics and Technology. His scientific work was mainly based on optical physics (around 100 publications) (Fig. 1.5).

One of Vavilov's collaborators, M. A. Konstantinova-Shlezinger, was developing methods for quantitative fluorescence analysis (in the luminescence laboratory of the Physics Institute of the USSR Academy of Sciences). She developed, in particular, the mentioned fluorescence methods for determining ozone in high layers of the atmosphere. She is the author of a monograph on fluorescence analysis [12] and was the compiling editor of another book under the same title [13]. Under the initiative of Vavilov she compiled and published abstract collections on fluorescence analysis and organized training in fluorescence analysis for employees of production laboratories.

Together with Vavilov, V. L. Levshin also carried out work on fluorescence. He studied polarized fluorescence and discovered the non-linear absorption of intense light flows by solutions of uranium compounds. Between 1931 and 1937 he established the rule of mirror symmetry of the absorption and fluorescence spectra (Levshin's rule). In 1934 he established the recombination character of the fluorescence of crystal phosphors; paid special attention to the study of the processes of intermolecular and intramolecular interaction in condensed solutions of various organic compounds, primarily dyes (1927–1968); and he created a theory of various concentration effects which developed with a significant decrease in intermolecular distances. He also published a monograph *Photoluminescence of liquid and solid substances*, which was published in the USSR, Hungary (1956), and China (1958). It was the most complete overview of work in this field.

A significant trace in the history of fluorescence analysis was left by E. V. Shpol'skii. In 1952, he, together with co-workers A. A. Il'ina and L. A. Klimova, discovered the appearance of quasilinear spectra of solutions of complex organic molecules in *n*-paraffin matrices at low temperatures [14, 15]. The phenomenon was called the "Shpol'skii effect," later it formed the basis for numerous procedures of fluorescence analysis (Fig. 1.6).

Fig. 1.6 Eduard Vladimirovich Shpol'skii (December 10(23), 1892–August 21, 1975), an optical physicist who discovered the "Shpol'skii effect" which is widely used in luminescence analysis. With permission of the Russian Academy of Sciences

***Eduard Vladimirovich Shpol'skii** (1892–1975) was a doctor of physical and mathematical sciences and a professor at the V. I. Lenin Moscow State Pedagogical Institute (MSPI). The scientific work of E. V. Shpol'skii dealt with spectroscopy, biophysics, photochemistry, and the history of physics. Between 1948 and 1950 he studied the spectra of polycyclic aromatic hydrocarbons (PAH), and in 1950–1951 he began to study the spectra of frozen solutions. In 1918, together with S. I. Vavilov and P. N. Lazarev, he started the journal Uspekhi Fizicheskikh Nauk [Advances in Physical Sciences], with which he was engaged for 56 years, since 1920, as editor in chief. He was also the editor of two journals Contemporary Problems of Natural Science and The Newest Trends of Scientific Thought. Since the foundation of the All-Union Institute of Scientific and Technical Information (in 1953) he was the chief editor of the abstract journal Physics (for 22 years).*

Research in this field was carried out at the Department of Theoretical Physics of MSPI (by T. N. Bolotnikova, E. A. Girdzhiyauskaite, R. I. Personov, and R. N. Nurmukhametov) and the Institute of Physical Problems of the USSR Academy of Sciences in the laboratory of P. L. Kapitsa (by L. A. Klimova).

A great contribution to the investigation of the Shpol'skii effect, and its uses, was made by R. I. Personov, who worked after leaving MSPI at the Institute of Spectroscopy of the USSR Academy of Sciences. His research included selective excitation of fine-structure fluorescence and phosphorescence spectra, analytical applications of the Shpol'skii effect, and the spectroscopy of single molecules. This work received great fame. The method of selective excitation of transitions in frozen organic matrices developed by R. I. Personov's group was used to solve a number of analytical problems, e.g., record high sensitivity for the determination of 3,4-benzpyrene was demonstrated. In 1961 the work of K. K. Rebane (Tartu) appeared, also shedding light on the nature of the Shpol'skii effect; the observed effect was proposed to be considered as an optical analog of the Mössbauer effect. In 1968 I. S. Osad'ko created a semiphenomenological theory of the vibrational structure of the absorption and fluorescence spectra of complex molecules.

The Shpol'skii method is used to study subtle effects associated with intramolecular and intermolecular interactions, transfer of excitation energy, association of molecules, etc. In Moscow, Shpol'skii's spectra were studied at the L. Ya. Karpov Physicochemical Institute, the V. I. Vernadskii Institute of Geochemistry and Analytical Chemistry of the USSR Academy of Sciences (GEOKHI), the All-Union Cancer Center, at the Department of Geography at Moscow State University (MSU), and at the Institute of Labor Protection. G. I. Romanovskaya described in detail the history of the discovery of the Shpol'skii effect [16].

A. N. Terenin (1896–1967) explained (in 1943) the triplet nature of the phosphorescent state of organic compounds. Terenin's book entitled *Photonics of Molecules of Dyes and Related Organic Compounds* (1967) was highly appreciated. A significant contribution to the development of the physical foundations of fluorescence was also made by B. I. Stepanov, N. A. Borisevich, and others, with Stepanov authoring several books on fluorescence (1955, 1963).

Among the Russian professional analytical chemists who made great contributions to the development and practical uses of fluorescence methods, one should note E. A. Bozhevol'nov, D. P. Shcherbov, I. A. Blyum, A. P. Golovina, K. P. Stolyarov, and N. N. Grigor'ev.

E. A. Bozhevol'nov sought opportunities for a variety of uses of the method, and in many ways contributed to making it a reliable tool for analytical chemistry. He studied the reasons why some molecules fluoresce and others do not. The explanation of this fact is the presence or absence of non-radiative transitions between the isoenergetic sublevels of a molecule. This assumption was tested on a large number of organic substances; the results made it possible to predict the luminescent properties of some compounds. He synthesized and proposed a number of organic substances as fluorescent reagents. He created a theory of organic fluorescent reagents based on the elimination of internal non-radiative transitions in the formation of complex compounds, and proposed sensitive reactions with their application. He developed fluorescence and kinetic fluorescent methods for the determination of microquantities of cations and organic compounds using low temperatures. The chemiluminescent methods developed in his work have become widespread, in particular, for the determination of ozone and nitrogen dioxide. Closely connected to numerous scientific centers and creating a school of fluorescence analysis, E. A. Bozhevol'nov always strived to ensure that the results of research became the property of analytical practice. His research monograph entitled *Fluorescence Analysis of Inorganic Substances* [17] was widely known to analysts.

Evgenii Aleksandrovich Bozhevol'nov *(1916–1975) was a doctor of chemical sciences, a professor, and laureate of the State Prize of the USSR (1972). From 1948 he worked at the All-Union Research Institute of Chemical Reagents and High Purity Chemical Substances (IREA). He authored more than 300 scientific papers, headed up a large research team, and was a member of the editorial board of the Journal of Analytical Chemistry as well as being on the editorial boards of other journals.*

Many extraction–photometric and extraction–fluorescent methods, using basic dyes, were developed by I. A. Blyum (mainly for the analysis of geological objects). He authored a book entitled *Extraction–Photometric Methods of Analysis Using Basic Dyes* [18]. D. P. Shcherbov, like I. A. Blyum, worked in the field of photometric and fluorescent analysis of mineral raw materials. He created many methods for the determination of various elements, developed analytical principles of photometry, introduced the concept of color saturation of colored solutions, developed narrow-band liquid filters, and produced nomograms that made it easier to calculate the results of an analysis. He developed methods for the fluorimetric determination of beryllium, rhenium, zirconium, indium, gallium, mercury, and silver. He was involved in the creation of the FO-1 fluorimeter produced by Geologorazvedka. He authored *Fluorimetry in Chemical Analysis of Mineral Raw Materials* [19].

Dmitrii Pavlovich Shcherbov *(1906–1981) was a doctor of chemistry, Honored Chemist of the Kazakh SSR, and ran the laboratory of photometric methods of analytical chemistry of the Kazakh Institute of Mineral Raw Materials (KazIMS). Before that, he was head of the physicochemical laboratory at the Lenfil'm film studio. During World War II he worked in the Ural analytical laboratories of the geological service. From 1946 he worked at Alma-Ata in the Central Laboratory of the South Kazakhstan Geological Administration as a senior chemist, then from 1956 he worked at KazIMS (in charge of a laboratory). In 1962 he became a doctor of chemical sciences. He had over 200 works published and actively participated in many conferences.*

A scientific group within the Department of Analytical Chemistry at M. V. Lomonosov MSU should also be noted (A. P. Golovina, N. B. Zorov, V. K. Runov, and others). In addition, it is worth mentioning the book written by A. P. Golovina and L. V. Levshin [20] as well as work by A. P. Golovina and co-workers—a large review entitled *Chemical Luminescence Analysis of Inorganic Substances* [21].

Alla Petrovna Golovina *(1925–2007) was a candidate of chemistry and associate professor at the Department of Analytical Chemistry at M. V. Lomonosov MSU. She created a group, from which several doctors of sciences (N. B. Zorov, V. K. Runov, S. V. Kachin, and others) emerged. Together with L. V. Levshin (at the Department of Physics at MSU) she investigated the luminescent spectroscopic properties of various dyes, which were widely used for analytical purposes, and actively developed fluorescent methods for determining the ions of elements. Golovina and Levshin wrote a textbook entitled Chemical Fluorescent Analysis of Substances which was published in 1978. The book advocated the use of fluorescent methods in analytical practices.*

V. K. Runov worked in the field of optical analytical spectroscopy (using fluorescence, colorimetry, and other methods) applied to the determination of small quantities of substances. He proposed selective fluorescent methods for the determination of platinum metals, copper, and other elements. He developed optical sorption-molecular-spectroscopic methods for the analysis of liquids and gases (the latter in conjunction with the NPO "Khimavtomatika"). He also participated in the development of spectral instruments.

Valentin Konstantinovich Runov *(1949–1999) was a doctor of chemical sciences, a professor, and was in charge of the laboratory of spectroscopic methods at the Department of Analytical Chemistry at M. V. Lomonosov MSU. He graduated, in 1971, from the faculty of chemistry at MSU, and subsequently worked at the Department of Analytical Chemistry. In 1994 he defended his doctoral thesis. He authored 170 publications and was involved with 40 inventions. He was a lecturer for a number of special courses for MSU students.*

At Leningrad University, work on luminescence analysis of inorganic substances was successfully carried out by K. P. Stolyarov and N. N. Grigor'ev [22]. For a review of recent papers on luminescence analysis, see the collective monograph [23].

1.4 Enhanced Two-Jet Arc Plasmatron and Other Work on Atomic Emission Analysis

In 1930, B. A. Lomakin proposed an empirical dependence of the intensity of spectral lines on the concentration of an element in a sample (*Proceedings of VNII Metrologii, 1932, Volume 2(18), pp. 139–163*). This dependence is known as the Lomakin–Scheibe equation (or Scheibe–Lomakin equation).

Much has been done in the direction of improving the two-jet arc plasmatron as a source of excitation of a spectrum. The source is convenient for the direct analysis of solid bulk samples, e.g., crushed geological or powdered industrial samples. In solving such problems, a two-jet arc plasmatron has advantages over other sources used for atomic emission analysis (Fig. 1.7).

The idea of a two-jet plasmatron, put forward and realized for the first time, apparently, by Valente and Schrenk [24], was developed by V. S. Engelsht and co-workers [25–28] and at the same time by Elliot et al. [29, 30]. The developments of Engelsht, Zheenbaev, and others, perfected by A. P. Tagiltsev, were first actively used by analysts in Novosibirsk under the leadership of I. G. Yudelevich [31], and then S. B. Zayakina. VMK-Optoelectronics, a firm in Novosibirsk, created atomic emission spectrometers with improved two-jet plasmatrons (e.g., [32]). A lot of experience in the practical use of a two-jet arc plasmatron, for the analysis of geologic samples and bulk materials, was accumulated and summarized in the doctoral thesis of S. B. Zayakina as well as in a monograph written by her in collaboration with G. N. Anoshin [33, 34] (Fig. 1.8).

For arc atomic emission analysis, the so-called "spill" method was developed (by V. V. Nedler and A. K. Rusanov). In a horizontally placed arc, a finely divided powder of the sample is introduced from above. The method is semiquantitative, but proved to be very convenient for the large-scale testing of mineral raw materials and was very widely used, especially in the geological service [35].

Since the end of the last century, VMK-Optoelectronics has been working to create and improve a linear multi-channel analyzer of emission spectra (MAES) based on multi-crystal assemblies of photodiode lines. MAES has found wide

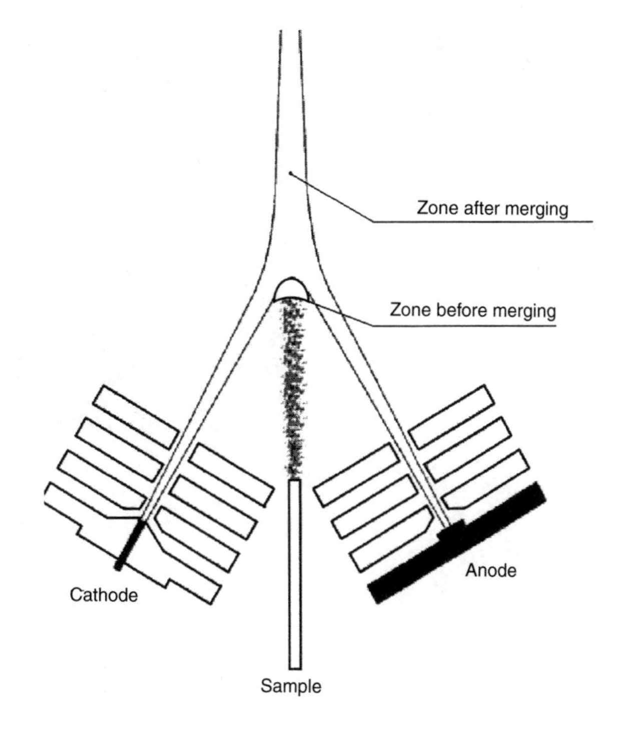

Fig. 1.7 The scheme of a two-jet plasmatron. Figure provided by S. B. Zayakina

application in analytical laboratories; it is used with various sources of excitation of radiation, i.e., an arc of direct and alternating current, spark, laser, inductively coupled plasma, two-jet arc plasmatron, along with various spectral devices, i.e., prism and diffraction, domestic and foreign. In 2001, it was included in the State Register of Measuring Instruments of the Russian Federation. MAES is continuously being improved on the basis of experience of use in numerous analytical laboratories.

1.5 X-Ray Polycapillary Optics

In the early 1980s an employee of the I. V. Kurchatov Institute of Atomic Energy, M. A. Kumakhov, proposed the principle of focusing X-rays, based on the use of polycapillary optics [36–39]. Later, he set up the Institute for Roentgen Optics (IRO). The principle of focusing was developed and practically implemented in devices of various types with various purposes.

It is well known that optical fibers are used for the optical range. They include a central "core" of transparent, optically dense material that is inside a flexible tube made from a material that is optically less dense. When a light beam hits the end of the light guide (at a certain angle), complete internal reflection takes place, and the light travels along the bent fiber, without leaving the limits of the central core. This

Fig. 1.8 Book by G. N. Anoshin and S. B. Zayakina about a two-jet plasmatron

is due to the fact that in the optical range, for practically all media, the refractive index of light is greater than unity. X-rays will not travel along such a light guide due to their strong interaction with matter and rapid damping, and also because the X-ray cannot be retained inside the light guide: in the X-ray range, the refractive index of the absolute majority of media is slightly less than unity. However, for the same reason, the X-ray may be retained in the hollow, slightly curved capillary. When a ray hits the capillary at a very small critical angle, the effect of total external reflection appears—the X-ray without interaction with the wall material is repeatedly reflected from the internal surfaces of the hollow capillary, without large attenuation passing through a less dense medium (air or vacuum).

A specially organized system, with axial symmetry of hundreds of thousands or millions of such capillaries, differing from each other by their radius of curvature and the nature of the change in the internal diameter of a single channel along the axis, may be used as an X-ray lens. Using such lenses, it is possible to obtain a parallel beam of X-rays and, most importantly, to focus X-rays. By focusing radiation from very weak sources, very high X-ray densities may be achieved at the focus of the lens. Such densities are achievable only on modern synchrotron accelerators of enormous power and volume (Fig. 1.9).

After the creation of the first capillary lens (in 1984), the associated technology was constantly improved. The first such lens consisted of 2000 channels with a diameter of about 0.5 mm, whereas a lens from the most recent generation consists of several million channels, of micron and submicron diameters. The main research on polycapillary optics and its applications was carried out at the abovementioned institute. At the same location, production of polycapillary lenses was organized, along with the development of analytical instruments using such technology.

Polycapillary optics have a very large angle of radiation capture—0.1 radians or more, compatible with simple and cheap X-ray sources—X-ray tubes. Optics make it possible to diverge rays from the radiation source into a quasi-parallel beam in both directions (X-ray half-lenses) or to focus the beam (lenses). Polycapillary lenses and semi-lenses have small dimensions (from a few millimeters to a dozen centimeters depending on their purpose); they can work in both a vacuum and air.

The possibilities of focusing X-rays determines the merits of the devices created on the basis of polycapillary optics: very low–power consumption (from a few watts to several tens of watts, instead of traditional kilowatt sources); light weight with small dimensions; and having multifunctionality. These devices do not require special rooms and protection; many of them may be used in the field; and they are convenient and easy to operate (Fig. 1.10).

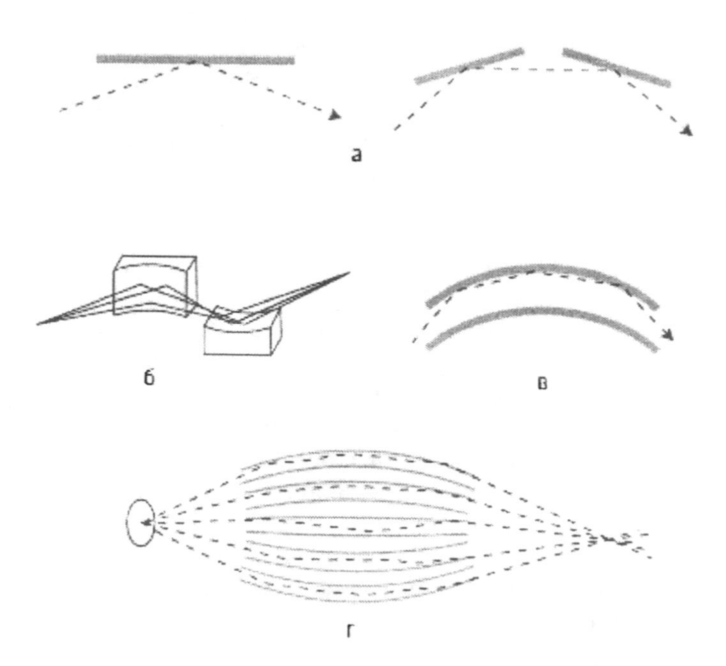

Fig. 1.9 The scheme of functioning of the Kumakhov lens. Figure provided by the Institute of X-Ray Optics

Fig. 1.10 Muradin Abubekirovich Kumakhov (July 29, 1928–June 25, 2014), a physicist and the creator of Kumakhov X-ray optics. Figure provided by the Institute of X-Ray Optics

A series of analytical instruments based on polycapillary optics have been developed with their own specific applications. Among these are multifunctional devices, e.g., the Mini-Lab-6 complex which combines a number of functions—elemental analysis, diffractometry, refractometry, etc. The X-ray fluorescent microanalyzer know as Focus-M (a spectrometer that makes it possible to carry out local elemental analysis) is used in many laboratories, including forensic centers. When using two lenses that look at one point, a three-dimensional image is obtained. A hand-held device has also been developed—a microanalyzer (weighing about 3 kg) for the analysis of rocks, alloys, art objects, and jewelry (Fig. 1.11).

One very interesting instrument is the laboratory synchrotron. This small device (weighing about 10 kg) produces a stream of monochromatic quasi-parallel beams comparable to those of middle-generation synchrotrons. This became possible owing to the creation of new technology for polycapillary lenses and a bright microfocus source. On the basis of polycapillary technology, an X-ray fluorescence analyzer was created with a sensitivity of $n \cdot 10^{-7}\%$. This device makes it possible to analyze water, solutions, etc.

There are about 1000 devices with Kumakhov optics which are used to solve various practical problems. Manufacturers of X-ray equipment—EDAX, Philips, Siemens, Bede, Shimadzu, Unisantis and other corporations—produce instruments with these optics. Such devices are used in large synchrotron centers to focus synchrotron radiation, as well as to obtain three-dimensional images of an object using two lenses. When focusing, the synchrotron radiation density increases by more than three orders of magnitude. Kumakhov's optics, and the devices based on such optics, have received numerous prizes and diplomas at international exhibitions and conferences.

The creation of polycapillary optics and corresponding instruments is undoubtedly one of the most important achievements in analytical X-ray technology in recent decades.

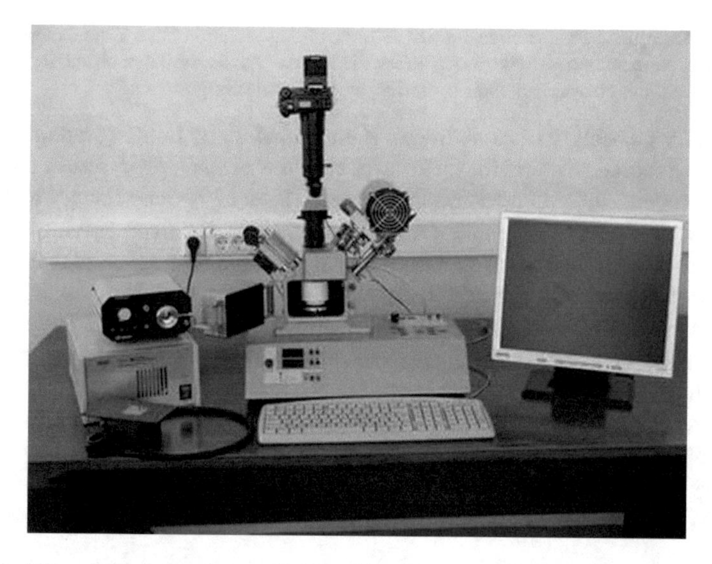

Fig. 1.11 "Focus-M" microanalyzer with Kumakhov X-ray lens. Figure provided by the Institute of X-Ray Optics

1.6 X-Ray Radiometric Analysis and Its Applications

This method of elemental analysis is based on the excitation of atoms of the elements to be determined using sources of ionizing radiation (usually radioactive) and analysis of the spectral composition of the resulting X-ray radiation of the excited atoms using radiometric equipment. Identification of the elements is carried out by their characteristic X-ray radiation, and this radiation is detected not by crystal-diffraction analyzers, as in ordinary X-ray fluorescence analysis, but, as already mentioned, using radiometric equipment. Usually, the energy-dispersive method of registration is used.

Here is the characteristic of the method given after many years of development and use [40]:

> "As compared with the X-ray tube, the radionuclide source has several advantages: (1) both photons for electromagnetic radiation (gamma- or X-ray) and heavy charged particles (e.g., α-particles) can be used to excite the atoms of the detected elements, (2) high stability of the source (in fact, when the half-life of the applied radionuclide is several months, for a sample measurement time of several minutes the source of the exciting radiation can be considered perfectly stable), (3) monochromatic nature of the excitation radiation which makes it possible to reduce the background in the analysis and to reduce the influence of the material composition of the samples on the results of the determinations, (4) the simple possibility of carrying out the analysis by harder and more intense lines of the K-series of the characteristic radiation, including heavy elements, (5) energy efficiency and compactness of the source.
>
> To the full, these advantages of radionuclide sources can be realized in compact energy-efficient equipment intended for testing ores in their natural occurrence (without sampling), for the analysis of samples in field laboratories, for well logging, etc.

At the same time, radionuclide sources of the exciting radiation have a number of disadvantages in comparison with X-ray tubes. The main one is the "non-switching" of the sources, which complicates their operation, storage, and transportation."

The very possibility of excitation of X-ray radiation by ionizing radiation from a radioactive source was previously known, e.g., it was sounded in one of the reports at the first Geneva Conference on the Peaceful Uses of Atomic Energy in 1955. In the USSR this idea was embodied in a method that involved solving numerous methodological and technical problems and at the same time developing equipment. The practical goal was the search for minerals and the analysis of mineral raw materials. Initially, work was carried out mainly at the All-Union Institute of Mineral Raw Materials (VIMS) in Moscow under the leadership of Alexander Lazarevich Yakubovich (Fig. 1.12).

Subsequently, A. L. Yakubovich was awarded the Prize of the Scientific Council of the Russian Academy of Sciences (RAS) on Analytical Chemistry, was an honorary member of this Council, and received many other awards. He fought in World War II, and went on to live a long and rich life. He died in 2014 at the age of 94.

Various instruments based on this method were produced, and are still produced, in Russia in large batches and are widely used by the geological service, at mining enterprises, and in other areas. The method has also been used in space research.

X-ray radiometric analysis and its various applications are described in detail in many books, (e.g. [40–44]).

Fig. 1.12 Aleksandr Lazarevich Yakubovich (October 30, 1919–May 22, 2014), creator of the methodologies and instrumentation for the X-ray radiometric search for uranium; much has been done to develop such X-ray radiometric and nuclear-physical methods of analysis. Photo provided by the All-Russian Institute of Mineral Raw Materials

1.7 Other Spectroscopic Methods

I. I. Kanonnikov (1854–1902), while at Kazan University, developed in 1880 a refractometric method for the analysis of organic compounds [45].

The method of atomic-absorption determination of mercury, using its "cold vapor" (a flameless method), was substantially improved. The foundations of the method were laid down by the work of Müller and Pringsheim [46], Woodson [47], and others. Poluektov, Vitkun, and Zelyukova [48, 49] raised the possibility of determining mercury, after its reduction in solution using $SnCl_2$, by air blowing atomic mercury into a special cell. What was new here was the use of tin (IV) chloride as a reducing agent (this has become widely used) and, in contrast to Woodson's work on the determination of mercury in air, it was proposed to recover mercury in solution and to blow out atomic mercury for its subsequent determination. It was after the work of Poluektov and his co-workers that the method of "cold vapor" became widespread.

Between 1970 and 1990, methods of laser analytical spectroscopy were rapidly developing in the world. Russian scientists made a significant contribution to this area of analytical chemistry. In V. S. Letokhov's group (at the Institute of Spectroscopy of the USSR Academy of Sciences), a variant of laser atomic photoionization spectroscopy (LAPIS), with the thermal evaporation of the sample, was developed, an appropriate instrument was designed, and the low detection limits of elements were realized. The analytical capabilities of the method were demonstrated, in particular, in the determination of low concentrations of rare-earth elements in sea water (Fig. 1.13).

In the work of M. A. Bol'shov's group, at the same institute, a method of laser atomic-fluorescence spectroscopy (LAFS), with a graphite atomizer, was developed —incorporating the theoretical foundations of the method along with the creation of unique instrumentation. Using a LAFS-1 spectrometer, placed in a clean room, heavy metal (Pb, Bi, and Cd) determinations were made in deep ice cores from

Fig. 1.13 Vladilen Stepanovich Letokhov (November 10, 1939–March 21, 2009), a physicist-spectroscopist and one of the pioneers of laser physics, including analytical laser spectroscopy. Photo provided by the Institute of Spectroscopy of the RAS

Antarctica and fresh snow from Greenland and the mountain groups of the Alps. The data obtained proved to be valuable for the determination of natural climate variations between about 200,000 years BC to present day. Natural variations of the concentrations of metals in prehistoric epochs (from 0.4 ppt of lead in warm climatic periods to 20 ppt in glacial periods) were found (Fig. 1.14).

At the same time, interesting work was being performed by N. B. Zorov's group (at Moscow University) to improve the method of laser-induced ionization in a flame. High sensitivity of the method was demonstrated, however, a significant dependence of the analytical signal on the composition of the sample was also revealed.

A. I. Nadezhdinskii et al., at the A. M. Prokhorov General Physics Institute of the RAS actively developed diode-laser spectroscopy for the purposes of chemical analysis. Nadezhdinskii regularly organizes conferences concentrating on this method. Many applications have been developed, e.g., for the determination of various gases, including applications linked to medical diagnostics [50].

Considerable research has been carried out in the field of thermo-optics, especially thermal lenses spectrometry (formerly undertaken by V. I. Grishko, V. P. Grishko, and I. G. Yudelevich [51], and continued presently by M. A. Proskurnin et al., at M. V. Lomonosov MSU) [52–54]. This is one of the most sensitive methods of molecular absorption spectrometry, although it is not widely used in practice, at least in Russia. Its scientific solutions are very interesting, including when this method is combined with others, and also when it is used in new analytical systems (microfluidics, etc.).

In the USSR, several groups worked on the use of electron spin (paramagnetic) resonance (discovered in 1944 by E. K. Zavoiskii in Kazan) as a method of chemical analysis [55] (Fig. 1.15). In general, methods were developed for the

Fig. 1.14 Institute of Spectroscopy of the Russian Academy of Sciences in Troitsk. Photo provided by the Institute of Spectroscopy of the RAS

Fig. 1.15 Evgenii Konstantinovich Zavoiskii [15(28)0.09.1907–09.10.1976] discovered electron paramagnetic resonance. Photo previously published

determination of paramagnetic ions of copper (II), molybdenum (V), manganese (II), chromium (III), vanadium (IV), iron (III), and some rare-earth elements. Portable ESR spectrometers have been created. A series of studies on spin-labeled analytical reagents (extractants), such as 1 and 2, has been performed.

O•N, N, C(=O)–CH_2–C(=O)–R

O•N, NH–C(=O), COOH

As a label in the molecules of more or less ordinary complexing reagents, stable nitroxide radicals were introduced. After complexing with metal ions (in this case also diamagnetic) and separating the excess of the reagent, if present, the metal concentration can be determined by measuring the ESR signal [56].

References

1. L'vov, B.V.: Zh. Anal. Khim. **60**, 134 (2005)
2. Walsh, A.: Spectrochim. Acta **7**, 108 (1955)

3. L'vov, B.V.: Inzhenerno-fizich. zhurnal, **11**, 44 (1959)
4. L'vov, B.V.: Inzhenerno-fizich. zhurnal, **11**, 56 (1959)
5. L'vov, B.V.: Spectrochim. Acta **17**, 761 (1961)
6. L'vov, B.V.: Atomic Absorption Spectrochemical Analysis, 392p. Nauka, Moscow (1966) (in Russian)
7. L'vov, B.V.: Atomic Absorption Spectrochemical Analysis, 324p. Adam Hilger, London (1970)
8. Massmann, H.: Spurenanalyse mittels Atomabsorption in der Graphitcüvetten nach L'vov. In: II International Symposium „Reinstoffe in Wissenschaft und Technik" (Dresden, 1965), 297p. Akademie-Verlag, Berlin (1965)
9. L'vov, B.V.: Spectrochim. Acta, Part B **33**, 153 (1978)
10. Vavilov, S.I.: Izv. Akad. Nauk SSSR, Ser. Fiz. **9**, 283 (1945)
11. Vavilov, S. I., Microstructure of Light. Izd. Akad. Nauk SSSR, Moscow (1950) (in Russian)
12. Konstantinova-Schlezinger, M.A.: Luminescence Analysis, 288p. Izd. Akad. Nauk SSSR, Moscow–Leningrad (1948) (in Russian)
13. Konstantinova-Schlezinger, M.A.: The Abstract Collection on Luminescent Analysis, Issue 1, Moscow, 1951; Issue 2, Moscow, 1954 (in Russian)
14. Shpol'skii, E.V., Il'ina, A.A., Klimova, L.A.: Dokl. Akad. Nauk SSSR, **87**, 935 (1952)
15. Shpol'skii, E.V.: Usp. Fiz. Nauk, **68**, 51 (1959)
16. Romanovskaya, T.I.: Zh. Anal. Khim. **64**, 775 (2009)
17. Bozhevol'nov, E.A., Luminescent Analysis of Inorganic Substances, 415p. Khimiya, Moscow (1966) (in Russian)
18. Blum, I.A.: Extraction-Photometric Methods of Analysis using Basic Dyes. Nauka, Moscow (1970) (in Russian)
19. Shcherbov, D.P., Fluorimetry in Chemical Analysis of Mineral Raw Materials. Brief Methodological Guideline, 260p. Nedra, Moscow (1965) (in Russian)
20. Golovina, A.P., Levshin, L.V.: Chemical Luminescence Analysis of Substances, Khimiya, Moscow (1978) (in Russian)
21. Golovina, A.P., Runov, V.K., Zorov, N.B.: Chemical luminescence analysis of inorganic substances. In: Structure and Bonding, pp. 47, 53. Springer, Berlin (1981)
22. Stolyarov, K.P., Grigor'ev, N.N.: Introduction to Luminescence Analysis of Inorganic Substances, 364p. Khimiya, Leningrad (1967) (in Russian)
23. Romanovskaya, G.I. (ed.): Luminescence Analysis (Problems of Analytical Chemistry, 20), Nauka, Moscow (2015) (in Russian)
24. Valente, S.F., Schrenk, W.G.: Appl. Spectrosc. **24**, 197 (1970)
25. Konavko, R.I., Engel'sht, V.S., Burnachiev, D., et al.: All-Union Conference on Generators of Low-Temperature Plasma, 155p. Ilim, Frunze (1974) (in Russian)
26. Engel'sht, V.S., Urmanbekov, K.U., Zheenbaev, Zh.Zh.: Zavodsk. Lab. 42 (1976)
27. Zheenbaev, Zh.Zh., Engel'sht, V.S.: Two-Jet Plasmatron, 202p. Ilim, Frunze (1983) (in Russian)
28. Tagil'tsev, A.P.: Spectral and Chemico-Spectral Methods for Analyzing Samples of Complex Composition using a Two-Jet Arc Plasmatron, Diss. Cand. Tech. Sciences, 66p. Frunze (1985) (in Russian)
29. Elliot, W.G., Karlinski, T.J.: Plasmajet Owice and Method of Operating Same. USA Patent No. 4.009.413, priority as of February 22, 1977
30. Spectrometrics, Incorporated Karl J. Hildebrant, Respublique Francias Patent. No. 7900753, priority as of January 12, 1979
31. Yudelevich, I.G., Cherevko, A.S., Tagil'tsev, A.P.: Izv. Sib. Otd. Akad. Nauk SSSR, Ser. Khim. Nauk **2**, 80 (1981)
32. Gerasimov, V.A., Labusov, V.A., Saushkin, M.S.: Two-Jet Arc Plasmatron for Atomic-Emission Spectral Analysis, RF Patent No. 2298889
33. Zayakina, S. B., Anoshin, G.N.: Two-Jet Arc Plasmatron in Analytical Spectrometry, 268p. LAP Lambert Acad. Publ., 2013 (in Russian)

34. Zayakina, S.B.: Two-Jet Arc Plasmatron in Atomic Emission Analysis of Geological Samples and Disperse Process Materials. Diss. Doct. Tech. Sciences, 351p. Novosibirsk, 2009 (in Russian)
35. Rusanov, A.K.: Fundamentals of Quantitative Spectral Analysis of Ores and Minerals, 2nd ed., revised and addit, 400p. Nedra, Moscow (1978) (in Russian)
36. Kumakhov M.A. USSR Inventor's Certificate No. 1322888, priority as of July 26, 1984
37. Kumakhov, M.A.: X-Ray Spectrom. **29**, 343 (2000)
38. Kumakhov, M.A.: Nucl. Instrum. Methods Phys. Res. B **48**, 283 (1990)
39. Nikitina, S.V., Shcherbakov, A.S., Ibraimov, N.S.: Rev. Sci. Instrum. **7**, 1 (1999)
40. Yakubovich, A.L., Ryabkin, V.K.: In: Yakubovich, A.L. (ed.) Nuclear-Physical Methods of Analysis and Quality Control of Mineral Raw Materials, 206p. VIMS, Moscow (2007) (in Russian)
41. Yakubovich, A.L., Zaitsev, E.I., Przhiyalgovskii, S.M.: Nuclear Physics Methods of Rock Analysis, 3rd ed., 263p. Energoizdat, Moscow (1982) (in Russian)
42. Malikonyan, S.V.: Equipment and Methods of Fluorescent X-ray Radiometric Analysis, 280p. Atomizdat, Moscow (1976) (in Russian)
43. Ochkur, A.P., Tomskii, I.V., Yanshevskii, Yu.P., et al.: In: Ochkur, A.P. (eds.) X-ray Radiometric Method for Prospecting and Exploration of Ore Deposits, 256p, Nedra. Leningr. otd., Leningrad (1985) (in Russian)
44. Plotnikov, R.I., Pshenichnyi, G.A.: Fluorescence X-ray Radiometric Analysis, 254p. Atomizdat, Moscow (1973) (in Russian)
45. Kanonnikov, I.I.: On the Relationship between the Composition and the Refractive Power of Chemical Compounds. Tipogr. University, Kazan (1883) (in Russian)
46. Müller, K., Pringsheim, P.: Naturwissensch, 364 (1930)
47. Woodson, Th.T: Rev. Sci. Instr. **10**, 308 (1939)
48. Poluektov, N.S., Vitkun, R.A.: Zh. Anal. Khim. **18**, 37 (1963)
49. Vitkun, R.A., Poluektov, N.S., Zelyukova, Yu.V: Zh. Anal. Khim. **29**, 691 (1974)
50. Stepanov, E.V.: Diode Laser Spectroscopy and Analysis of Biomarker Molecules, 416p. Fizmatlit, Moscow (2009) (in Russian)
51. Grishko, V.I., Grishko, V.P., Yudelevich, I.G.: Laser Analytical Thermal Lens Spectroscopy, 322p. Novosibirsk (1982) (in Russian)
52. Proskurnin, M.A., Kononets, M.Yu.: Usp. Khim. **73**, 1235 (2004)
53. Proskurnin, M.A., Volkov, D.S., Gor'kova, T.A., Bendrysheva, S.N., Smirnova, A.P., Nedosekin, D.A.: Zh. Anal. Khim. **70**, 227 (2015)
54. Proskurnin, M.A.: Zh. Anal. Khim. **71**, 451 (2016)
55. Solozhenkin, P.M.: Electronic Paramagnetic Resonance in the Analysis of Substances, 292p. Donnish, Dushanbe (1986) (in Russian)
56. Nagy, V.Yu., Petrukhin, O.M., Zolotov, Yu.A: Crit. Rev. Anal. Chem. **17**, 265 (1987)

Chapter 2
Mass Spectrometry and Related Methods

Abstract Mass reflectron developed by B. A. Mamyrin is widely used in time-of-flight mass spectrometers, as well as the orthogonal entry proposed by A. F. Dodonov. Mass spectrometers of the "Orbitrap" series, developed by A. A. Makarov and co-workers at Thermo Co., were widely disseminated and recognized. Electrospray ionization was proposed by L. N. Gall and co-workers before the work of J. Fenn, who received a Nobel Prize for Chemistry. Ion–molecule reactions, identified by V. L. Talrose, opened the way to chemical ionization. Russian specialists contributed to secondary-ion mass spectrometry and fast atom bombardment, proposed photoionization at atmospheric pressure, and mass spectrometry of negative ions formed as a result of resonance electron capture.

2.1 General Remarks

For the last two to three decades, mass spectrometry has become one of the most important and common methods of chemical analysis. It incorporates a number of versions that differ in their areas of application (isotopic, elemental, and molecular analysis; working with labile bioorganic substances) and technical characteristics. Technical and design distinctions are very significant and include versions that differ in mass analyzers, ionization and registration methods, or techniques for introducing a substance into the ion source.

Speaking of mass analyzers, Russian specialists have made a significant contribution to time-of-flight mass spectrometry (TOFMS). An essential place is occupied here by the electron mirror (mass reflectron) created by B. A. Mamyrin, and further improved by A. N. Verenchikov. The orthogonal entry of a continuous ion beam, developed by A. F. Dodonov and co-workers, was a great achievement.

As for ionization methods, many have been proposed, though with different scales of use and practical significance. Electron ionization still remains the most utilized method; it used to be the only one for producing ions at the early stages of mass spectrometry. Among other methods, we can mention spark ionization, secondary-ion ionization, fast atom bombardment (FAB), inductively coupled

Y. A. Zolotov, *Russian Contributions to Analytical Chemistry*,
https://doi.org/10.1007/978-3-319-98791-0_2

plasma ionization, chemical ionization, photoionization, surface ionization, electrospray ionization (ESI), and MALDI (Matrix Assisted Laser Desorption and Ionization). Open-air ionization techniques have been intensively developed recently; quite a few are already available.

Some of the ionization methods were proposed in the Soviet Union (Russia). V. L. Talrose and A. K. Ljubimova (1952) created the notion of ion–molecule reactions, concepts that paved the way for chemical ionization. Formation of negative ions by resonance electron capture (REC) was a result of the work of V. I. Khvostenko and co-workers. I. A. Revelsky and Yu. S. Yashin (1983) were the first to propose photoionization under atmospheric pressure. It was in the Soviet Union that the pioneering research on ESI was carried out by L. N. Gall and co-workers in 1980–1983, although it is commonly believed that the technique was actually created by John Fenn, who was one of the Nobel Laureates for Chemistry in 2002. The same can be said about FAB: G. D. Tantsyrev published his work before Michael Barber and Alfred Benninghofen, who are usually given credit for the development of this method.

There are several groups in modern Russia that are active in analytical mass spectrometry. These groups are headed by A. A. Sysoev, E. N. Nikolaev, L. N. Gall, I. A. Revelsky, A. T. Lebedev, V. G. Zaikin, E. S. Brodskii, etc. In addition, there is an All-Russia Mass-Spectrometric Society as well as the journal *Mass Spectrometry*.

The first mass spectrometers were created by Joseph Thomson in 1912, Arthur Dempster in 1918, and Francis Aston in 1919. The world's fourth mass spectrometer was manufactured by V. N. Kondratev at the Institute of Chemical Physics under the supervision of N. N. Semenov. This was to become the dissertation of a future academician Kondratev.

2.2 Mass Reflectron

As a result of long-term experiments that started in 1965 and lasted for no less than 25 years at the Ioffe Physical–Technical Institute of the USSR Academy of Sciences, B. A. Mamyrin created a new type of TOFMS, which he coined the mass reflectron [1, 2].

A TOFMS is based on the fact that ions move at different speeds in an electric field depending on their masses or, to be precise, their mass-to-charge ratios. All ions formed in an ion source move, and this determines the unlimited mass range. The device is also characterized by a high-speed response. Such analyzers are simple from the technical point of view. However, they also have some limitations, namely, they cannot operate with continuous ion sources, since all the ions must start from a certain point. In addition, ions have certain non-zero and differently directed velocities at an initial time instant, and they fly after gaining acceleration in the form of a somewhat diffused packet. This leads to an unsatisfactory separation of peaks and a low resolution. Creation of a mass reflectron, to a certain degree, solved this problem by compensating the spread in the kinetic energy of ions.

The electron mirror, called the reflectron, was invented by B. A. Mamyrin, V. I. Karataev, D. V. Shmikk, and V. A. Zalulin in the late 1960s–early 1970s [1, 2]. The principle of its operation was concisely described by specialists as follows [3]:

> "The reflector of ions, or an ion mirror, which is a system of grid electrodes, gave the name 'mass reflectron' to the entire device. In this device, higher-energy ions with a certain mass penetrate more deeply into a reflecting electric field, thereby spending more time on retardation and subsequent acceleration. While being the first to reach the ion mirror, they are the last to leave it and catch up with the slower ions of the same mass by the time of reaching a detector (which is usually the assembly of two microchannel plates). Thus, a difference in the time of flight in a field-free region (before and after the reflector) is compensated for by the dwell time in the reflector area. A time focusing of the same-mass ions that fly with different velocities thus occurs, and this is how the resolving ability of the analyzer is improved. One more advantage of reflectron is its much smaller dimensions as compared to linear time-of-flight devices, in which a maximum-length tube is required for effective mass separation of ions…"

Here is an excerpt from an interview with B. A. Mamyrin from 2006 [3]:

> "I received the inventor's certificate for mass reflectron back in 1967 [4]; I also patented it in the UK, Japan, France, and the USA [5]. They waited until the patent support expired and immediately started to produce reflectron-based mass spectrometers."

Mamyrin's devices, incorporating a mass reflectron, have the following important advantages: an unlimited range of recorded masses, the ability to observe the entire mass spectrum, and high sensitivity. Time-of-flight spectrometers of this type are currently produced by many companies and are used extensively. For example, they have been used for human genome sequencing (Fig. 2.1).

Apart from creating a mass reflectron, Mamyrin also carried out important research on helium isotopes [6] and the determination of fundamental physical constants [7, 8]. He and his group created magnetic resonance mass spectrometers with record high sensitivities ($\sim 3.10^4$ helium atoms per analyzer volume) for isotopic analysis. The devices were manufactured at the Institute for Analytical Instrumentation of the USSR Academy of Sciences under MI 9301, MI 9302, and other trademarks. The FTIAN-series mass spectrometers have been used at metallurgical plants for the continuous online monitoring of steelmaking (Fig. 2.2).

Boris Aleksandrovich Mamyrin *(May 25, 1919–March 5, 2007) was born on May 25, 1919, in Lipetsk, into a family of medical doctors. In 1941, he graduated from the Department of Physics and Mechanics of Leningrad Polytechnic Institute as a physicist. He saw active service in the Soviet–Finnish War in 1939–1940 and World War II, and he was awarded two orders and many medals. Demobilized in 1948, he came back as a disabled war veteran and started work at the Ioffe Physical–Technical Institute of the USSR Academy of Sciences in Leningrad. From 1970, he first headed a division at the laboratory of mass spectrometry and, then, from 1983, the laboratory itself. In 1949, he defended his candidate of sciences thesis entitled "Modulating devices for an installation for separating uranium isotopes by the high-frequency method," and in 1966, he defended his doctoral thesis entitled "Studies in the area of time-of-flight ion separation." In 1994, he was elected a corresponding member of the Russian Academy of Sciences. In 2000,*

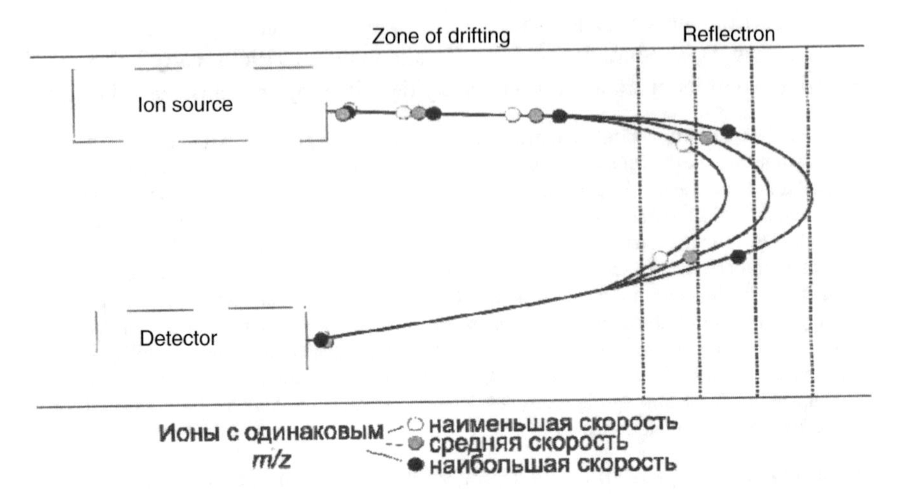

Fig. 2.1 Principle scheme of a mass reflectron. With permission of A.T. Lebedev

Fig. 2.2 Boris Aleksandrovich Mamyrin, the creator of the mass reflectron. Provided by B.A. Mamyrin in 2005.

the American Society for Mass Spectrometry awarded him "The Distinguished Contribution in Mass Spectrometry" Award. Mamyrin became laureate of the academician B. P. Konstantinov Award of the USSR Academy of Sciences for the series of work comprising "Research and development of a mass-spectrometric method for metal-production monitoring." He was also a member of the international Committee on Data for Science and Technology (CODATA). In his later years, he was the chief scientific officer in a mass spectrometry laboratory. He died in St. Petersburg on March 5, 2007.

Here is what Mamyrin wrote to the author of this book about the mass reflectron on August 11, 2005.

“II. Mass reflectron (its description and diagram are available in the proof prints).

Information about it was published for the first time in Soviet Union Inventor’s Certificate no. 198034 (*Byul. Izobret.*, 1969, no. 13, p. 148). It was patented in the USA, the UK, France, and Germany. However … our government was not supporting foreign patents with all the consequences that came with it.

Time-of-flight mass spectrometers are different from static magnetic ones by an unbounded mass range and high analysis speed. However, they had a low resolution of $R \leq 300$. The use of the reflection of ions in an electrostatic mirror made it possible to compensate for the effect of ion-energy spread that is inevitable whichever ion formation mechanism is used. (When reflected in a retarding field, the “fast” ions penetrate more deeply and thus arrive at the detector simultaneously with the “slow” ones.)

Our first publication (Mamyrin et al.) in *Zh. Tekh. Fiz.*, 1973, vol. 37, p. 45 on the spectrum of heavy organometallic molecules for $R \geq 6000$ quickly swept across the world. Many companies started to develop mass reflectrons. Devices with $R \approx 10\,000$ were developed approximately 10 years later. Attached find a photograph with the internal design of one such device manufactured by the Bruker company. It accommodates 100 samples at once, with analysis of one sample taking only a few seconds.

The American scientist Craig Venter set up a whole “workshop” of such devices; this made it possible to quickly complete the human genome sequencing (10^5 genes!). An international program on the human proteome ($>10^6$ cells![1]) has been started recently; this will make it possible to dramatically shorten the time needed to create “side-effect-free medicines”…

Small reflectrons are used in medicine, food quality checks, criminology, doping control, and so on. And, of course, it is applied in chemistry and physics. A lot of details and schemes are in the proof prints being sent.

We developed special-purpose automated mass-reflectron installations FTIAN-3, FTIAN-4, and FTIAN-5 for monitoring of metallurgical processes; these installations are now being produced by a plant in the city of Sumy.

Pneumatic steelmaking now operates without rejects or oxygen reblowing, both of which protract the processes, and, most importantly, without gas-duct explosions, which led to major accidents or even loss of human life. Two small companies seceded from the laboratory to produce compact mass-reflectron installations with inferior analytical parameters that suffice for some applications. Nearly all metallurgical plants in Russia, CIS countries, and certain other countries are equipped with FTIAN installations. They are also used to control various metallurgical processes.

Reflectrons find their use in monitoring the production of ultrapure inert gases such as hydrogen, ensuring a sensitivity to impurities of up to 10^{-7}–10^{-10}.”

Later, A. N. Verenchikov and co-workers developed a multireflection mass reflectron that included a series of electron mirrors and provided a resolution of 25,000–50,000. This mass spectrometer is produced serially in combination with gas or liquid chromatographs.

[1*] Rightly: proteins.—*Yu.A. Zolotov.*

2.3 Orthogonal Entry of a Continuous Ion Beam

In the late 1980s, Dodonov et al. [9, 10] developed a system for orthogonal entry of a continuous ion beam at the Institute for Energetical Problems of Chemical Physics of the USSR Academy of Sciences. The system was designed mainly as an electrospray source (the ERIAD technique) and made it possible to connect this source to a mass reflectron (Mamyrin's TOFMS with an electron mirror). This considerably enhanced the capabilities of the mass reflectron. The invention had a tremendous impact and quickly found applications (Fig. 2.3).

Aleksandr Fedorovich Dodonov *(January 18, 1939–August 31, 2005) graduated from the Moscow Physico–Technical Institute in 1966. He first worked at the Institute of Chemical Physics of the USSR Academy of Sciences, and later, from 1987, at the Institute for Energetical Problems of Chemical Physics at the same academy. His research interests were in mass spectrometry, to which he contributed a number of fundamental works. He is best known for his contribution to TOFMS, viz., the development of the orthogonal entry of ions with the use of different ionization sources, including the use of electrospraying.*

Soon after, foreign colleagues proposed a similar entry scheme for a source based on electron ionization, too.

Nowadays, instrument-making companies actively use orthogonal entry.

Fig. 2.3 Aleksandr Fedorovich Dodonov proposed an orthogonal input for time-of-flight mass spectrometers. Provided by the Institute of Energetic Problems of Chemical Physics

2.4 Orbitrap: Orbital Ion Trap

The very much appreciated Orbitrap, orbital ion trap, was created by Makarov [11], who worked at Thermo Co. in Germany. To advocate the importance of including this achievement within this book, let us look at two excerpts from an interview with A. A. Makarov made by the authors of [12] (Fig. 2.4).

> "Question: The first and foremost question is related to how you and your team-mates approached the creation of the Orbitrap. Of special interest is what it started with, whether it happened gradually or at once, how it all came about and was developed further.
>
> Answer: Everything started when I was a student at the Moscow Engineering and Physics Institute (1984–1989), where A. A. Sysoev had a postgraduate student, Aleksandr Leopol'dovich Pekal', who once told me: 'Since you want to work here anyway, why don't you invent an ideal mass spectrometer that could solve my problem: a spark ion source with huge angular and energy spreads?' It was then that I began to study fields with quadratic focusing and came across Golikov's and other works. As a result, my Candidate of Sciences dissertation was devoted to fields with ideal focusing: 'Isochronous motion of charged particles in static electromagnetic fields.'
>
> Later, in 1992, when I defended my thesis and became a Candidate of Mathematical and Physical Sciences, I took a trip, kind of a chapiteau circus tour across Britain, Belgium, and Germany; however, I did not get far. I made it only as far as Manchester, where I was literally caught by people from Kratos Analytical Ltd. The group included people who later set up our small company, of which the core were people from Russia—Eduard Denisov, Aleksandr Kholomeev, and myself.

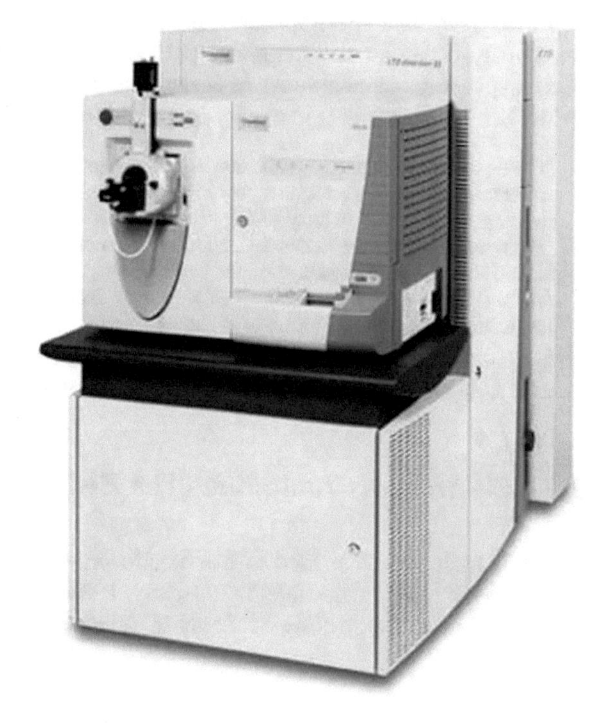

Fig. 2.4 Orbitrap mass spectrometer. Taken from the advertising brochure of Thermo Scientific Co

Question: Can you share with us how difficult the problems related to electronics and hardware were in this difficult work on developing a fundamentally new device?

Answer: Yes, indeed, we essentially had to invent something totally new in order to achieve the parameters we needed. The stability of the electronics was of great importance, as it governed the stability of the entire device, especially in its commercial version. This was not of utmost importance for research, but crucial for a commercial device. Therefore, credit for the fact that we managed to quickly transition from a research to a commercial version goes to the specialists and, first of all, to Aleksandr Kholomeev.

Question: Do these people still work with you?

Answer: Yes.

Question: So, were it not for the three Russians, nothing would have happened?

Answer: Yes. The very method appeared because of the existing time-of-flight mass-spectrometry schools in St. Petersburg and Moscow. This was already the launch pad from which we started. Of course, nothing would have happened if we did not create this company or protect it from the things that were happening in the harsh struggle for survival under 'wild capitalism.' Of course, without a doubt, credit for this goes to them, including support with the electronics and mechanics and useful discussions—all of this is extremely important. Russian mass spectrometrists definitely constituted the core of the group.

Question: It would be good to hear you say that 'domestic brains' were engaged in the creation of such a unique device.

Answer: This is precisely what is clear from this story! Were it not for these three specialists or if at least one of them had not been there, there would have been no device. Surely, something could have been made, but this would have happened much later."

In the new mass spectrometer, a linear ion trap is combined with the Orbitrap mass analyzer via a radio frequency (RF) quadrupole bent-axis ion trap [13]. The mass analyzer developed by A. A. Makarov and his team does use magnetic or RF fields, but is an electrostatic axially-harmonic orbital ion trap.

"Owing to the fact that axial oscillation is independent of the kinetic energy of ions and that an electric field can be set with a very high precision and stability, the Orbitrap is an ideal mass spectrometer by its very principle of operation. This implies that an ultrahigh resolution can be achieved with this device and the ion mass can be measured very accurately. It is no small matter that in actuality, the large capacity of ions proved typical of this device, thus making it possible to attain highly accurate mass measurements, a wider dynamic range, and a greater range of mass-to-charge ratios [13, 14]."

Specialists now recall with a smile that at the time, mass-spectrometry connoisseurs had formulated 12 reasons why an Orbitrap-type trap could not be made.

2.5 Electrospray Ionization (The ERIAD Technique)

In 1968, Malcolm Dole tried to use electrospraying in combination with a mass spectrometer [15]. In the late 1970s–early 1980s, at the Institute for Analytical Instrumentation of the USSR Academy of Sciences, Lidiya Nikolaevna Gall, along

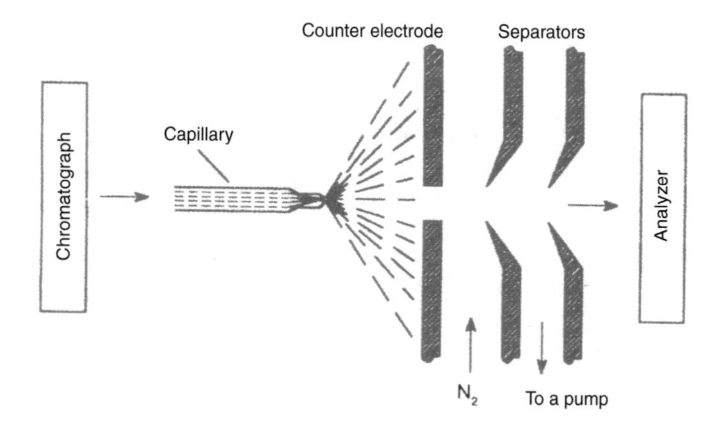

Fig. 2.5 Principle scheme of an electrospray. With permission of A.T. Lebedev

with the group she supervised, developed a method they named ERIAD (the Russian acronym for the "extraction and spraying of ions under atmospheric pressure"). The method was highly rated and even became the subject of articles in popular science journals (Fig. 2.5).

The method is essentially as follows (according to [16]). A syringe pump is used to direct the flow from liquid chromatograph, or from a capillary tube, into a 0.1-mm-diameter needle that is fed a voltage of the order of several kilovolts. An aerosol of charged droplets with a high surface charge is formed at the output from the needle. The droplets move toward a grounded counter electrode. The pressure drops in the same direction, although, on the whole, it is kept equal to atmospheric pressure in this part of the ion source (before the counter electrode). Moving toward the exit orifice (or the capillary tube) of the first separator, the droplets shrink due to solvent evaporation. When they reach a critical size, when the surface-tension forces can no longer resist the Coulomb repulsion forces (the Rayleigh limit), the droplet "explodes" to form finer droplets. This process is repeated all over again, resulting in microdroplets, each containing just one charged particle that can find itself in a gas phase after the residual solvent molecules evaporate. Heating or collisions with the molecules of an inert gas can be used for ultimate desolvation (Fig. 2.6).

Below is a description of the history of the development of ESI taken from [16] (Figs. 2.7 and 2.8).

> "A certain contribution to the theory and creation of electrospraying was made by Dole [15]. The first ESI experiments led by Gall started in Leningrad in 1979. The method they created was called ERIAD, the Russian acronym for the 'extraction and spraying of ions under atmospheric pressure'. The method was realized on a static sector device with double focusing, and the first spectra of bradykinin, multicharged insulin ions, and involatile salts were obtained in 1981. It was shown that application of a high voltage difference

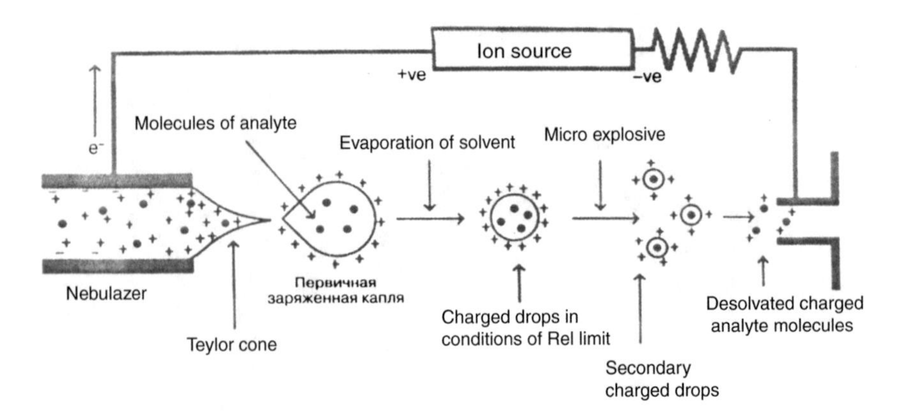

Fig. 2.6 Sequential loss of shell by ions during electrospray. With permission of A.T. Lebedev

Fig. 2.7 Lidiya Nikolayevna Gall (born September 1, 1934). Together with co-workers, she created the ERIAD method (electrospray) and developed a whole series of mass spectrometers. *Courtesy* of L.N. Gall

(~ 1000 V) between the output end of the electrospray capillary and the skimmer can initiate fragmentation of protonated molecules without collisional activation. The results were reported at numerous meetings in the Soviet Union in the early 1980s. Unfortunately, the closed nature of Soviet science prevented Gall from publishing her results in top-rated international journals. What is more, even in Russian, the results were published very late, in April 1984 [17, 18], because most leaders in domestic mass spectrometry were wary of her method. John Fenn, however, expressed interest in the results when he visited the Soviet Union in 1983. The fate of Fenn's research proved much more fortunate. Although his first work was published only in September 1984 [19, 20], that is, after Gall's pioneering work, he managed to unlock the potential of electrospraying for the international scientific community by later publishing the numerous results of his own experiments. Fenn was awarded the Nobel Prize in 2002 for the development of ESI."

Since then, electrospraying has found the widest possible range of applications.

Доклады Академии наук СССР
1984. Том 277, № 2

УДК 543.51 543.8 + 54.07 ФИЗИЧЕСКАЯ ХИМИЯ

М.Л. АЛЕКСАНДРОВ, Л.Н. ГАЛЛЬ, Н.В. КРАСНОВ,
В.И. НИКОЛАЕВ, В.А. ПАВЛЕНКО, В.А. ШКУРОВ

ЭКСТРАКЦИЯ ИОНОВ ИЗ РАСТВОРОВ ПРИ АТМОСФЕРНОМ ДАВЛЕНИИ – МЕТОД МАСС-СПЕКТРОМЕТРИЧЕСКОГО АНАЛИЗА БИООРГАНИЧЕСКИХ ВЕЩЕСТВ

(Представлено академиком Е.П. Велиховым 28 IX 1983)

Основные преимущества масс-спектрометрии, связанные с высокой чувствительностью метода, возможностью определения молекулярной массы и получения структурной информации, практически не используются при анализе труднолетучих, полярных и термонестабильных соединений, к которым относится большинство веществ биоорганического происхождения. Это обусловлено невозможностью испарить их без разложения при обычном нагревании образца. Необходимо

Fig. 2.8 The first article on electrospraying

2.6 Ion–Molecular Reactions as a Path to Chemical Ionization

In 1952, Talrose and Ljubimova [21] observed a signal from methonium ions CH_5^+ in an electron ionization source under elevated methane pressure in an ionization chamber. This observation was later deemed rather important; it is no coincidence that many years later, in 1998, the *Journal of Mass Spectrometry* reprinted the article [21] as one of great historical significance [22].

In 1966, Manson and Field [23] created a suitable source and used chemical ionization for analytical purposes. Ionization with ion–molecular reactions has become widely accepted as an ionization method [24] (Fig. 2.9).

Victor L'vovich Talrose *(April 15, 1922–June 22, 2004) graduated from Moscow State University (MSU) in 1947. He saw active service in World War II. In 1947–1948, he worked at the Institute of Chemical Physics of the USSR Academy of Sciences, and in 1986, became the director of the Institute for Energetical Problems of Chemical Physics of the USSR Academy of Sciences, which he created. At the same time, he was also teaching at the Moscow Physico–Technical Institute, where he became a professor in 1961 and was the dean of the department of molecular and chemical physics. Talrose held a doctoral degree in chemistry and was a corresponding member of the USSR (Russian) Academy of Sciences. He was a physicist and chemist, with his main work being devoted to the kinetics of chemical reactions. Talrose conducted a lot of research in mass spectrometry, including reactions of free ion radicals. In 1969, he created the first chemical laser based on*

Fig. 2.9 Victor L'vovich Talroze. Provided by the Institute of Energetic Problems of Chemical Physics

the reaction between hydrogen and fluorine. Talrose led the development of Russia's first chromato-mass-spectrometers. He did much to develop scientific instrument making, headed the Council on Scientific Instrument Making of the USSR Academy of Sciences, and chaired the Committee on Mass Spectrometry of the USSR Academy of Sciences. Talrose was awarded government decorations. In the last years of his life, he spent a lot of time in the United States.

A book of memoirs [25] was published about Talrose. One book [26], devoted to those who contributed most to mass spectrometry, contains a chapter about him.

The history of the concept of chemical ionization was retold in detail by the renowned Russian mass spectrometrist E. N. Nikolaev (note that the text was split into two paragraphs without the author's approval).

"His first work, which V. L. Talrose started, most likely, at the instigation of Viktor Nikolaevich Kondratev, who headed the laboratory Talrose had just joined at that time; it dealt with the study of electron impact ionization of methane. In this study, he encountered the very interesting phenomenon of the formation of an unusual CH_5^+ ion consisting of one carbon and five hydrogen atoms, a composition that did not conform with the existing notions of carbon valence. By that time, it had already been established that carbon valence is four and surely not five. This result was published in the journal *Dokl. USSR Acad. Sci.*, in an article (V. L. Talrose and A. K. Ljubimova, *Dokl. Akad. Nauk SSSR, 1952, vol. 86, p. 909*) submitted to the editorial board by Lev Davidovich Landau, an outstanding Soviet theoretician who had a better understanding of what valence means than anybody, but still he presented the article. Nobody believed the result claimed in the article, and the ultimate proof for the existence of such an ion was only obtained with a double focusing mass spectrometer, which was then considered a device with ultrahigh resolution: only one existed at the Institute for Physical Problems of the USSR Academy of Sciences. It was demonstrated that the ions formed are indeed CH_5^+ ions rather than OH^+ as many believed. Perhaps nobody would have remembered this were it not for the work by the American chemists Field and Munson, who studied ionization of methane and other saturated hydrocarbons and also discovered that ionization results in the formation of protonated molecules of saturated hydrocarbons, including CH_5^+. They announced the discovery of a new ionization method, which they called chemical ionization, which in their opinion essentially consisted in the following. If one adds a small amount of molecules of a substance *M* that one wants to identify into methane and ionizes the resultant mixture by electron impact, it is the methane molecules that will mostly be ionized rather than … those of substance *M*. Methane ionization will produce CH_5^+ ions that, when colliding with the molecules … of substance *M*, will donate a proton and thereby ionize the latter without destroying them, instead forming protonated molecules—quasimolecular ions. This method thus allows one to overcome the main difficulty of mass-spectrometric analysis: the destruction … of a molecule subjected to a direct electron impact (Fig. 2.10).

This discovery gave rise to an entire series of works devoted to chemical-ionization processes, so that some time after, chemical ionization firmly entered the practice of mass-spectrometric analysis and remains there. Many modern mass spectrometers are equipped with chemical ionization sources, while the mass-spectrometry community surely recognized the fact that V. L. Talrose was the first to obtain this result in 1952. Not without some struggle, Talrose's work was admitted as fundamental for this ionization method. It was highly commended while attracting a large number of citations. The CH_5^+ ion was named methonium, with its structure still debated—whether it is a complex of CH_3^+ with a hydrogen molecule or the so-called fluxonium ion, in which hydrogen atoms are constantly mixed. Graham Cooks (Purdue University), who is considered one of the most

Fig. 2.10 V.L. Talroze, Institute of Energetic Problems of Chemical Physics, Moscow. Provided by the Institute of Energetic Problems of Chemical Physics

well-known specialists in mass spectrometry, classified the discovery of chemical ionization as one of ten most significant results ever obtained in mass spectrometry in the twentieth century, while the name of V. L. Talrose was firmly imprinted in the history of modern mass spectrometry."

2.7 Secondary-Ion Mass Spectrometry and Fast Atom Bombardment

In the 1960s, at the Leningrad Polytechnic Institute, Mikhail Aleksandrovich Eremeev was probably one of the first to study and employ secondary-ion emission. Secondary-ion mass spectrometry (SIMS) was also developed at Khar'kov and other places, too.

The history of FAB leads us to the conclusion [27] that the method was simultaneously and, apparently, independently developed by several scientists, including Tantsyrev, Devienne, Benninghofen, and Barber. Tantsyrev published his first work [28–30] in 1971–1973, at about the same time as Benninghoven [31, 32] and Devienne. The method, however, was only recognized after the appearance of Barber's article [33] in 1981. This publication did not go unnoticed and gave impetus to the development of relevant devices. All of the authors mentioned above gave their own names to the method, but only the term "fast atom bombardment" (FAB), proposed by Barber, has been enshrined in scientific history. Barber is therefore considered to have developed the method, although Benninghofen is sometimes attributed with its development. However, as pointed out above, Tantsyrev was the first (Fig. 2.11).

In his 1985 review, Tantsyrev noted that in 1972, he and Nikolaev [29] proposed a mechanism for the formation of sprayed substances, based on the "concept of local heating of the surface of a target which now became a common one." He also

Fig. 2.11 Georgii Dmitrievich Tantsyrev made a great contribution to the creation and development of mass spectrometry of secondary ions and was the author of the first work on ionization by bombardment with fast atoms. Provided by the Institute of Energetic Problems of Chemical Physics

wrote that, again together with E. N. Nikolaev, he was "the first to discover the emission of proton-based ion clusters of the type $H^+ (H_2O)_n$ [for] $n \leq 50$ in ice bombardment experiments" [28] (quoted as per [34]). In the same review, Tantsyrev considered the following characteristics of the FAB method.

> "Fast atom bombardment can be used to ionize substances with any molecular mass. This method makes it possible to produce both fragment ions and quasimolecular positive ions. It can not only discriminate individual substances but also determine the composition of solid and liquid samples and, in some cases, even their supramolecular structure. Fast atom ionization (FAI) is relatively easy to implement on a mass spectrometer of any type."

V. A. Pokrovskii later wrote [27]:

"As far as we know, G. D. Tantsyrev did not bring up the issue of the priority of his results and, with his natural humility, assumed that his publications and inventor's certificates would suffice. Unfortunately, the role of his works has remained underappreciated in the modern scientific literature worldwide" (quoted according to [34]).

In the aforementioned interview [34], Tantsyrev was directly asked about the sequence of development of FAB.

"Question: Could you tell us, please, whether Barber admits that you were the first to propose FAI?

Answer: We did it prior to Barber, but he published his results immediately, without patenting them. We decided to first obtain an inventor's certificate, though, and this takes approximately a year and a half. As for our publications, they appeared almost simultaneously, independently of each other."

Georgii Dmitievich Tantsyrev *was born on August 15, 1929, in the village of Golitsyno in Penza Oblast. After finishing school, he entered the Bauman School in Moscow, studied there for 3 years, and was transferred to the Moscow Engineering Physics Institute, from where he graduated. Throughout his entire career, he worked at the Institute of Chemical Physics of the USSR Academy of Sciences, initially led by V. L. Talrose. In a team with V. L. Talrose and V. I. Gorshkov, he created a chromatographic mass spectrometer with a capillary column. G. D. Tantsyrev died in 2007. Views about G. D. Tantsyrev are given in* [34].

2.8 Atmospheric Pressure Photoionization

Mass spectrometry with atmospheric pressure photoionization was proposed and developed by Revelsky and Yashin [35, 36]. In this case, analytes were ionized by ultraviolet radiation with quanta that had energy below the water ionization potential (10.6 and 10.2 eV) and above the ionization potentials of most organic compounds. This made it possible to eliminate the registration of water clusters. In addition, the low energy of the radiation quanta and atmospheric pressure allowed them to minimize dissociative breakup of the formed excited molecular or quasi-molecular ion. Conditions were found which allow practically exclude the dissociative decomposition of these ions in atmospheric pressure photoionization (APPI) and atmospheric pressure photochemical ionization (APPCI) and the recording of cluster ions of the analytes.

The technique is significant because a molecular or quasi-molecular ion that allows one to judge the molecular mass of a compound is the most important ion in the mass spectrum of the compound. The intensity of peaks due to these ions in electron ionization mass spectra often turns out to be low, or in some cases there are no peaks at all. This intensity is higher in chemical ionization mass spectra, but the sensitivity is much lower in this case. In the case of chemical ionization under atmospheric pressure, the sensitivity is higher, but the mass spectrum is in many cases burdened by the presence of cluster ions of water and of the analytes themselves. This complicates the interpretation of mass spectra and reduces the sensitivity of recording peaks of molecular ions. Photoionization under atmospheric pressure solves this problem.

APPI and APPCI mass spectra have been studied for a large (over 200) number of compounds [37–39]. In all cases, the respective mass spectra consisted only of M^+ or $(MH)^+$ ions, depending on the nature of the compound and the vapors of the reagent used in the APPCI mass spectrometry. The limits of detection in the case of APPI gas chromatography–mass spectrometry (GCMS) were 10^{-12}–10^{-10} g in the single-ion detection mode and 10^{-14}–10^{-12} g in the case of APPCI in the same mode detecting mass spectra. Means for the analysis of multicomponent mixtures without their separation have been developed [39, 40]. An approach to determining the number of impurities in pure organic substances has been proposed that is based on combining chromadistillation (CD) and chromato-chromadistillation [39, 41, 42] with APPCI mass spectrometry and makes it possible to detect a significantly larger number of impurities as compared with GCMS.

APPI GCMS has been shown [39, 42] to be capable of more reliably detecting the number of known impurities of aromatic hydrocarbons in complex hydrocarbon mixtures such as gasoline. Fast and highly selective discrimination of phthalates, phosphates, and phosphonates, as well as polyaromatic hydrocarbons in complex mixtures with the use of short capillary columns in APPCI GCMS, has been demonstrated in [42], with detection limits of 10^{-13}–10^{-12} g. The developed method for detecting polyaromatic hydrocarbons [43] is the fastest and most selective of the known techniques for determining these compounds. The creators of the method combined mass spectrometry (APPI/APPCI) not only with capillary gas chromatography (GC) but also with microliquid chromatography [44], a fact that has been recognized [45]. To be fair, when analyzing the purity of compounds from the impurities that can, in principle, be analyzed by the GC method, the combination of GCMS and APPI (APPCI) is preferable due to the efficiency of the capillary columns and smaller noise compared with high-performance liquid chromatography (HPLC), as well as due to the higher sensitivity of GCMS APPCI.

Nevertheless, the use of mass spectrometry APPI (APPCI) combined with HPLC has become widespread when determining various compounds, especially thermally unstable and volatile ones. The second work devoted to HPLC/mass spectrometry (APPCI) appeared only 10 years after [44].

The GCMS APPI/APPCI method was used to create a new approach for checking the quality of chemical and pharmaceutical products and extra pure–grade organic substances (monomers, reference samples). The approach makes it possible to detect a significantly larger number of impurities than the standard approach based on GCMS with electron ionization.

I. A. Revelsky and Yu. S. Yashin's priority in creating APPI and APPCI mass spectrometry combined with gas chromatography is commonly acknowledged (Fig. 2.12).

Igor' Aleksandrovich Revelsky *was born on June 8, 1936. He graduated from the Military Academy of Chemical Warfare in 1960. He was a professor and held a doctoral degree in chemistry. Dr. Revelsky is the Leading Researcher in the Division of analytical chemistry of the Lomonosov MSU, Honored Chemist of the Russian Federation. Dr. Revelsky's interests included different versions of chromatography, chromato-mass spectrometry, supercritical fluid extraction, sorption preconcentration, discrimination of components in complex mixtures of organic compounds, determination of ultralow concentrations of ecotoxicants, and estimation of the grade of purity of organic compounds. He developed techniques for determining the molecular masses of compounds with gas-density detectors and the*

Fig. 2.12 Igor' Aleksandrovich Revelskii proposed photoionization and photochemical ionization at atmospheric pressure. Provided by I.A. Revelsky

cross sections of the ionization of molecules. He also confirmed the hypothesis about the additivity of molecular ionization cross sections. Dr. Revelsky designed a high-temperature (up to 800 °C) gas-density detector as well as microcoulometric and titrimetric detectors. He developed APPI and APPCI mass spectrometry, which made it possible to record mass spectra that only consist of a molecular or quasimolecular ion.

2.9 Resonance Electron Capture Negative Ion Mass Spectrometry

The idea of resonance electron capture negative ion mass spectrometry (RECNIMS) was conceived in the 1960s–1970s in Ufa. The method was developed by Viktor Ivanovich Khvostenko (1933–1996). Khvostenko started his research into negative ions in Leningrad, at the Physico-Technical Institute of the USSR Academy of Sciences under the guidance of Dukel'skii [46, 47]. He continued his research in 1959 at the Bashkirian Branch of the USSR Academy of Sciences in Ufa [48, 49]. The history of the creation and development of this method has been thoroughly covered in publications by scientists who continue to work in this area [50, 51].

The method is based on resonance attachment of electrons to molecules, a phenomenon that occurs in the energy range of 0–15 eV. The REC mass spectra are three-dimensional, the dimensions being the energy coordinate, the mass number (or the *m*/*z* ratio, to be precise), and the peak intensity. In this case, a mass spectrum is a set of effective yield curves for all types of formed negative ions. This method is the only version of mass spectrometry to produce three-dimensional spectra.

In principle, any mass spectrometer can be adapted to operating in this mode, but the adaptation is not very easy to make. The complexity of the method, and the lack of mass production of relevant devices or, at least, of attachments to commonly used mass spectrometers, implies that the technique failed to be widely disseminated despite an enormous amount of accumulated research material (approximately 400 publications, 8 doctoral dissertations). The method nevertheless still offers ample opportunities for application.

Let us quote from a synthesis report [51].

> "For example, REC NI MS proved its potential for discriminating isomers by the occurrence of new resonances or by their shift along the energy scale. Let us dwell on just one example. NI MS makes it possible to observe a peak from molecular ions M^- of ortho-carborane, while no M^- peak is observed from *meta*- and *para*-isomers all the way to 0.0001% (!), although fragment ions are approximately the same for all the three isomers. There are simply no precedents for such striking distinctions between isomers in traditional mass spectrometry."

2.10 In Pursuit of Ultrahigh Resolution

The development and multiple applications of ion cyclotron resonance combined with Fourier transform allowed E. N. Nikolaev to attain the unique resolving ability of mass spectrometry at the Institute of Energetical Problems of Chemical Physics. Here is how Nikolaev himself described this (the text was split into two paragraphs without the author's approval) [52] (Fig. 2.13).

> In the 2000s, the device was considerably reformed with the participation of M.V. Gorshkov it was equipped with an electrospray ionization source. This source made it possible to ionize large biological molecules such as peptides and proteins. Just like all the other laboratories conducting research in the area of ion cyclotron resonance mass spectrometry, we took part in studies devoted to biological mass spectrometry, primarily, proteomics. We significantly contributed to global mass spectrometry owing to the invention of a special type of trap for ion cyclotron resonance spectrometers. The invented new type of Penning trap—a measure cell in an ion cyclotron resonance spectrometer—made it possible to improve the resolution of devices by an order of magnitude. The idea was taken up by the Bruker company (a join patent) which now produces a new type of ion cyclotron resonance devices, Solarix, the resolving ability of which for a mass of 1000 can be as high as 10 000 000 in relatively low magnetic fields. We called these dynamic harmonization traps. In these traps, electric fields that act on ions quadratically depend on the distance to the center in all directions. In fact, such a dependence can be achieved only after averaging the trajectory of ions due to their cyclotron movement. Work on improving these types of traps continues, and new resolving-ability records are being broken.

Fig. 2.13 Evgenii Nikolaevich Nikolaev (born January 1, 1947) developed mass spectrometric methods with special resolving powers and is a specialist in ion cyclotron resonance. Provided by E.N. Nikolaev

We also revisited the old ideas of detecting an ICR signal using a multielectrode arrangement. We proposed these ideas back in mid-1980s and patented and experimentally confirmed them. By using such an idea, the resolving ability of an ICR mass spectrometer can be improved without increasing the magnetic field. The Bruker company have now started producing devices with doubled detection frequency, which ensures twice as high resolution as ordinary dipole detection. The resolution of 10,000,000 allows one to view the fine isotopic structure in mass spectra; this is a new coordinate that makes it possible to determine the atomic composition of ions being detected from the distribution of the intensities of peaks in the fine isotopic structure."

2.11 Ion Mobility Increment Spectrometry

The now widespread technique of ion mobility spectrometry (IMS) was initially called plasma chromatography [53–55]. In essence, this technique is close to both mass spectrometry and chromatography. This method is used for detecting explosives and poisonous agents, drugs, and other chemical compounds. A number of companies produce devices, especially portable ones, that operate based on this technique. The history of the method was depicted in detail by Eiceman and Karpas [56] (see also [57]). The underlying principles of the technique were formulated by McDaniel at the Georgia Institute of Technology in the1950s–1960s.

The Russian contribution to this was the development of ion mobility increment spectrometry. This trend started with M. P. Gorshkov's patent [58]; the method was then being developed in Novosibirsk [59, 60]. Russian achievements in this area were reviewed by I. A. Buryakov in [61].

In the United States, this area is being actively explored by the Russian expatriate E. G. Nazarov.

References

1. Mamyrin, B.A., Karataev, V.I., Shmikk, D.V., Zagulin, V.A.: J. Exp. Theor. Phys. **37**, 45 (1973)
2. Mamyrin, B.: Int. J. Mass Spectrom. **206**, 251 (2001)
3. Khatymov, R.V., Khatymova, L.Z., Mazunov, V.A.: Istor. Nauki Tekh. **5**, 38 (2012)
4. Mamyrin, B.A., Time-of-flight mass spectrometer. USSR Inventor's Certificate No. 198034 (Byul. Izobret., 1969, **13**, 148) (January 14, 1966)
5. Mamyrin, B.A., Karataev, V.I., Shmikk, D.V.: Time-of-flight mass spectrometer. US Patent No. 4072862, 1978; UK Patent No. 1474149, 1977; French Patent No. 7530831, 1978; FRG Patent No. 2532552, 1980
6. Mamyrin, B.A., Tolstikhin, I.N.: Helium isotopes in nature. Ed. by: Fyle, W.S. Amsterdam: Elsevier Science (1984), 288 pp
7. Rapid Commun. Mass Spectrom., **21**, 1691 (2007)
8. Aleksandrov, E.V., Alferov, Zh.I., Anofriev, G.S., Ariev, N.N., Varshalovich, D.A., Zabrodskii, A.G., Kaplyanskii, A.A., Karataev, V.I., Mazets, E.P., Petrov, M.P., Usp. Phys. Nauk, 663 (2007)

9. Dodonov, A.F., Chernushevich, I.V., Dodonova, T.F. et al., A method of time-of-flight mass-spectrometric analysis from continuous ion beams. USSR Inventor's Certificate No. 1681340F1 (February 25, 1987)
10. Dodonov, A.F., Chernushevich, I.V., and Laiko, V.V., Time-of-flight mass spectrometry. In: Cotter, R.J. (ed.) ACS Symposium Series No. 549, 108 pp. Washington, DC: ACS (1994)
11. Makarov, A.A.: Anal. Chem. **72**, 1156 (2000)
12. Khatymova, L.Z., Mazunov, V.A., Khatymov, R.R., and Makarov, A.A., Current problems in the history of natural sciences in the areas of chemistry, chemical technologies, and oil industry. Proceedings XI International Conference. Ufa: Reaktiv, 227 pp. [in Russian] (2010)
13. Anon., Candidates in line to receive the "For Achievements in Mass Spectrometry" medal, Mass Spektrometry, **4**, 149 (2007)
14. Tokarev M.I., What is mass spectrometry and why is it needed? http://www.textronica.com./basic/ms_2.html
15. Dole, M., Heines, R.L., Mack, L.L., Mobley, R.C., Ferguson, L.D., Alice, M.B.: J. Chem. Phys. **49**, 2240 (1968)
16. Lebedev, T.A.: Mass spectrometry in organic chemistry, 2nd edn. Tekhnosfera, Moscow (2015) [in Russian]
17. Aleksandrov, M.L., Gall', L.N., Krasnov, N.V., Nikolaev, V.I., Shnurov, V.A.: Dokl. Akad. Nauk SSSR **277**, 379 (1984)
18. Aleksandrov, M.L., Gall', L.N.: Bioorg. Khim. **10**, 710 (1984)
19. Yamashita, M., Fenn, M.J.: Phys. Chem. **88**, 4451 (1984)
20. Yamashita, M., Fenn, M.J.: Phys. Chem. **88**, 4671 (1984)
21. Talrose, V.L., Ljubimova, A.K.: Dokl. Akad. Nauk SSSR **33**, 955 (1952)
22. Talrose, V.L., Ljubimova, A.K.: J. Mass Spectrom. **38**, 502 (1998)
23. Manson, M.S.B., Field, F.H.: J. Am. Chem. Soc. **1966**, 25 (1031)
24. Polyakova, A.A., Revelsky, I.A., Tokarev, M.N., Kogan, L.O., Talrose, V.L. et al., Mass-spectral analysis of mixtures with the use of ion–molecular reactions. In: Polyakova, A. A. (ed.) Moscow: Khimiya (1989) [in Russian]
25. Anon., Our Talrose: memoirs on the occasion of the 85th Anniversary of the Corresponding Member of the Russian Academy of Sciences V.L. Talrose. Moscow: Nauka (2007) [in Russian]
26. Gross, M.L., Carrioli, R.M. (eds.): The Encyclopedia of Mass Spectrometry, Vol. 10: Historical Perspectives. Part B. Notable People in Mass Spectrometry. In: Nier, K.A., Yorgey, A.L., Gale, P.J. (Vol. eds.) Amsterdam: Elsevier (2015)
27. Pokrovskii, V.A.: Fast atom desorption. In: Essays on the History of Mass-Spectrometry. Ufa: Bashkir Scientific Research Center, Urals Branch, USSR Academy of Science, 1988, 58 pp. [in Russian]
28. Tantsyrev, G.D., Nikolaev, E.N.: Pis'ma Zh. Eksp. Teor. Fiz. **13**, 473 (1971)
29. Tantsyrev, G.D., Nikolaev, E.N.: Dokl. Akad. Nauk SSSR **206**, 151 (1972)
30. Tantsyrev, G.D., Kleimenov, N.A.: Dokl. Akad. Nauk SSSR **213**, 649 (1973)
31. Benninghoven, A., Jaspers, D., Sichtermann, W.: Appl. Phys. **11**, 35 (1976)
32. Benninghoven, A.: Phys. Stat. Sol. **34**, K169 (1969)
33. Barber, M., Bordoli, R.S., Sedgwick, R.D., Tyler, A.N.: J. Chem. Soc. Chem. Commun., 325 (1981)
34. Khatymova, L.Z., Mazunov, V.A., Khatymov, R.V.: Istor. Nauki Tekh., **3**, Special Issue No. 1, 56 (2000)
35. Revelsky, I.A., Yashin, Yu.S., Voznesenskii, V.N., Kurochkin, V.K., Kostyanovskii, R.G.: A technique of mass-spectrum analysis of a gas mixture. USSR Inventor's Certificate No. 1159412 (Byul. Izobret., 1985, No. 47)
36. Revelsky, I.A., Yashin, Yu.S, Voznesenskii, V.N., Kurochkin, V.K., Kostyanovskii, R.G.: Izv. Akad. Nauk SSSR, Ser. Khim. **1986**, 9 (1987)
37. Revelsky, I.A., Yashin, Yu.S, Kurochkin, V.K., Kostyanovskii, R.G.: Ind. Lab. **57**, 243 (1991)

38. Revelsky, I.A., Yashin, Yu.S, Sobolevsky, T.G., Revelsky, A.I., Miller, B., Oriedo, V.: Eur. J. Mass Spectrom. **9**, 497 (2003)
39. Revelsky, I.A., Yashin, Yu.S: Talanta **102**, 110 (2006)
40. Revelsky, I.A., Yashin, Yu.S, Kurochkin, V.K., Kostyanovskii, R.G.: Zavodsk. Lab. **57**(3), 1 (1991)
41. Yashin, Yu.S, Revelsky, I.A., Tikhonova, I.N., Glazkov, I.N., Vulykh, P.P.: Zavodsk. Lab. **72**(11), 3 (2006)
42. Revelsky, I.A., Yashin, Yu.S, Zhukhovitskii, A.A.: Zavodsk. Lab. **56**(7), 24 (1990)
43. Revelsky, I.A., Tikhonova, I.N., Yashin, Yu.S: Eur. J. Mass Spectrom. **21**, 753 (2015)
44. Revelsky, I.A., Yashin, Yu.S., Tuulik, V.V., Kurochkin, V.K., Kostyanovskii, R.G.: Proceedings 1st All-Union Conference on Ion Chromatography. Moscow:, 43 pp. [in Russian] (1989)
45. Kersten, H., Kroll, K., Haberer, K., Brockmann, K.J., Benter, Th, Peterson, A., Makarov, A.: J. Am. Soc. Mass Spectrom. **27**, 607 (2016)
46. Khvostenko, V.I., Dukel'skii, V.M.: Zh. Eksp. Teor. Fiz., **33**, 851 (1957)
47. Khvostenko, V.I., Dukel'skii, V.M.: Zh. Eksp. Teor. Fiz., **34**, 1026 (1958)
48. Khvostenko, V.I.: In: Essays on the History of Mass-Spectrometry. Ufa: Bashkir Scientific Research Center, Urals Branch, USSR Academy of Science, 69 [in Russian] (1988)
49. Khvostenko, V.I.: Mass Spectrometry of Negative Ions in Organic Chemistry. Moscow: Nauka, 159 pp. [in Russian] (1981)
50. Khatymova, L.Z., Mazunov, V.A., Khatymov, R.V.: Istor. Nauki Tekh. **3**, 11 (2011)
51. Muftakhov, M.V., Khatymova, L.Z., Khatymov, R.V., Mazunov, V.A.: Izv. Bashkir Sci. Cent. Russ. Acad. Sci. **4**, 38 (2014)
52. Nikolaev, E.N.: Istor. Nauki Tekh. **3**, 38 (2017)
53. Cohen, M.J., Karasek, F.W.: J. Chromatogr. Sci. **8**, 330 (1970)
54. Karasek, F.W.: Anal. Chem. **46**, 710A (1974)
55. Lubman, D.M., Kronic, K.M.N.: Anal. Chem. **54**, 1546 (1982)
56. Eiceman, G.A., Karpas, Z., Hill Jr., H.H.: Ion Mobility Spectrometry, 3rd ed., 444 pp. CRC Press, Boca Raton (2016)
57. Wilkins, C.L., Trimpin, S. (eds.): Ion Mobility Spectrometry-Mass Spectrometry. Theory and Applications. CRC Press, Boca Raton (2010)
58. Gorshkov, M.P.: A method for analysis of impurities in gases. USSR Inventor's Certificate No. 966583, MKI G01 No. 27/62, 1982
59. Buryakov, I.A., Krylov, E.V., Soldatov, V.P.: A method for analysis of microimpurities in gases. USSR Inventor's Certificate No. 1485808, MKI G01 No. 27/62, 1989
60. Buryakov, I.A., Krylov, E.V., Makas', A.L., Nazarov, E.G., Pervukhin, V.V., Rasulev, U. Kh.: Pis'ma Zh. Tekh. Fiz. **17**, 60 (1991)
61. Buryakov, I.A.: Zh. Anal. Khim. **66**, 1210 (2011)

Chapter 3
Chromatographic Methods

Abstract Chromatography itself was developed by the Russian biochemist M. S. Tswett in the early 20th century. The first article on thin-layer chromatography was published in 1938 by N. A. Izmailov and M. S. Schreiber. N. M. Turkeltaub and M. M. Senyavin were among the first who developed gas chromatography before the known works of A. Martin; A. A. Zhukhovitskii, A. V. Kiselev, and V. G. Berezkin also made a significant contribution to this method. The original chromatomembrane method was developed by L. N. Moskvin and co-workers. V. A. Davankov authored the enantioselective ligand-exchange chromatography. Of interest is the critical chromatography of polymers (A. V. Gorshkov and V. V. Evreinov). A technology for obtaining polycapillary chromatographic columns was developed in Novosibirsk.

3.1 General Remarks

Although chromatography was developed by Russian botanist and biochemist M. S. Tswett at the very beginning of the 20th century, development before World War II was slow. There were, however, Kuhn's well-known work on the application of the method in biochemistry and Dubinin's work on the separation of gases, which were closely related (1935–1936); N. A. Izmailov and M. S. Schreiber, in 1938, published the first paper on thin-layer chromatography (TLC). Extensive research and, especially, analytical applications fall within the postwar period; at first in terms of ion exchange and gas chromatography.

In gas chromatography A. A. Zhukhovitskii and N. M. Turkeltaub proposed several original solutions in the 1950s–1960s; Turkeltaub's work on the separation of hydrocarbons was among the first in the world (1948–1950). V. A. Davankov, in the 1960s, initiated the concept of enantioselective ligand-exchange chromatography. A. V. Kiselev studied, and in 1962–1964 classified, adsorbents, going on to propose a number of new ones. B. G. Belenkii, E. S. Gankina et al. performed gel filtration in TLC. Chemists in Novosibirsk studied how to make polycapillary columns and went on to demonstrate their practical efficiency (1980s). L. N. Moskvin and co-workers in Leningrad (now St. Petersburg) invented liquid-gas chromatography and developed

Y. A. Zolotov, *Russian Contributions to Analytical Chemistry*,
https://doi.org/10.1007/978-3-319-98791-0_3

a chromatomembrane method (1980s–1990s). St. Petersburg experts also contributed to the development of monolithic chromatographic columns. O. A. Shpigun and co-workers are actively developing ion chromatography. There were advances in the development of instruments—the first microcolumn liquid chromatograph ("Ob" and "Milichrom"), various detectors for gas chromatography and high-performance liquid chromatography (HPLC), etc. A great number of applied problems were solved, e.g., in oil refining and petrochemistry, biotechnology and other fields.

In Russia, there is a powerful community of chromatographists. This community formed the Chromatography Commission of the USSR Academy of Sciences, which later transformed into the Scientific Council on Chromatography. This council carried out a significant amount of useful work in terms of convening conferences, including international conferences, an improvement of terminology, and publication of work. Attention was paid to Memorial activity, training of highly qualified personnel, and stimulation of development and production of devices. A review of Russia's achievements in the field of chromatography was given in the book entitled *100 Years of Chromatography* [1].

3.2 Birth of Chromatography

The birth of the chromatographic method is usually considered to be in 1903. Admittedly, initial development of this method was made by the botanist and biochemist M. S. Tswett, who worked in this direction first in St. Petersburg, then, for a longer time, in Warsaw, which was then part of the Russian Empire (Fig. 3.1).

M. S. Tswett separated vegetable pigments using a technique that today we would call liquid adsorption chromatography via elution. Calcium carbonate and other substances were used as the stationary phase. The first report on the new

Fig. 3.1 Mikhail Semenovich Tswett (May 14, 1872–June 26, 1919) was the primary developer of chromatography. Photo previously published

method was made by Tswett in 1903 [2] with his first journal publications dated 1905 and 1906 [3, 4] (Fig. 3.2).

Tswett had forerunners—the history of chromatography was considered in detail by E. M. Senchenkova [5]. In the 1940s–1950s there were attempts to challenge the primacy of M. S. Tswett, but discussions on this topic ended with recognition that it was Tswett who first considered and developed the chromatographic method. A lot has been written about the history of the development of the method; as an example, one can specify books by E. M. Senchenkova [5–7] or the publications of L. Ettre [8–10]. Therefore, there is no need to cover this history further in order to reconfirm M. S. Tswett's position. One can express sincere respect and deep gratitude to him (Fig. 3.3).

In 2013 the most complete summary of the works of M. S. Tswett was published [11], with extensive comments by the same E. M. Senchenkova, who gave several decades to the study of the history of chromatography.

In memory of M. S. Tswett conferences are held and a medal was established after him.

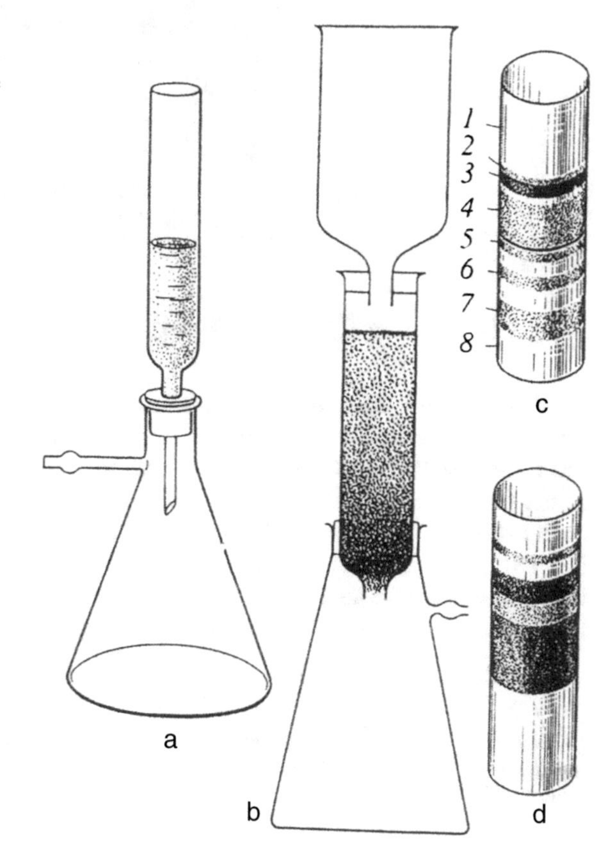

Fig. 3.2 M.S. Tswett's device for chromatographic analysis of **a** small and **b** large amounts of substances. **c** Chromatograms of the natural pigments of green leaves with their stratification into eight zones and **d** the same pigments treated with acid. Figure previously published

Fig. 3.3 A book published in 2013: M. S. Tswett's *Selected Works* (2013). Moscow: Nauka, 679 pp

3.3 Thin-Layer Chromatography

Professor E. Stahl, the author of the first, and very prestigious, book on TLC [12], a man who sometimes is considered to have pioneered the method, presented his book to the professor of the Khar'kov University N. A. Izmailov and a research associate, a pharmacist named M. S. Schreiber, with the dedication "I delegate my book to Prof. N. Izmailov and M. Shreiber, pioneers of thin-layer chromatography" [13, 14]. In the book Stahl points to the primacy of Izmailov and Schreiber. Their importance in the creation of TLC was recognized by many others. V. G. Berezkin [15] collected a number of statements on this topic:

> "Dr. Schreiber, together with Professor N. A. Izmailov, developed thin-layer chromatography. Their first publication on the technique appeared in 1938" [16].

> "Following column chromatography, came the development through the 1940s of paper chromatography as described by two Russian workers Izmailov and Schraiber in 1939" [17] (the year is mistakenly mentioned here).

"The first thin layer chromatogram was developed in 1938 by Izmailov and Schreiber in the circular mode" [18, 19].

"Bei der Suche nach Trenomoglichkeiten im Micromastab führten 1938 die beiden Russishen Wissenschaftler Izmailov und Schraiber die ersten versuche auf einer 'offenen Saule' 'also auf einer dunnen Schicht, durch'" [20].

"The technique of TLC was first used in 1937–1938 at the Institute of Experimental Pharmacy in Kharkov (Ukraine) by Nikolai A. Izmailov (1907–1961) and Maria Schraiber (1904–1992), his graduate student" [21] (Fig. 3.4).

ФАРМАЦИЯ

1938 № 3

ЛЕКАРСТВЕННЫЕ ПРЕПАРАТЫ

(Производство и испытание)

КАПЕЛЬНО-ХРОМАТОГРАФИЧЕСКИЙ МЕТОД АНАЛИЗА И ЕГО ПРИМЕНЕНИЕ В ФАРМАЦИИ

Сообщение I

И.А. Измайлов и *М.С. Шрайбер*

Физико-химическая лаборатория Украинского института экспериментальном фармации, Харьков

Основы хроматографического метода анализа впервые разработаны русским ботаником Цветом при исследовании растительных пигментов. Метод заключается в том, что исследуемый раствор смеси веществ пропускают через трубку, наполненную адсорбентом. Вещества при прохождении через адсорбент располагаются по зонам, в зависимости от адсорбционного потенциала. Для более полного распределения веществ применяют «проявление», т. е. через трубки пропускают ток чистого растворителя, который постепенно дифференцирует отдельные зоны. Таким образом, в трубке получают ряд зон, содержащих каждая одно определенное вещество или более простую смесь веществ. Получается так называемая хроматограмма.

Зоны, если они окрашены, обнаруживаются на дневном свету, а если они флуоресцируют – в ультрафиолетовом свету.

Результаты своих исследований Цвет опубликовал в 1910 г. (1). Однако в течение последующих лет метод Цвета мало применялся и только в 1931 г по-настоящему привлек к себе внимание исследователей, показавших, что хроматографический метод анализа может оказать большие услуги во многих областях исследования Подробный перечень работ о применении хроматографического метода анализа при исследовании растительных веществ, углеводородов, продуктов коксобензольного производства, пигментов, витаминов и др. можно найти в обзорном докладе Стикса (2), а также в работе Эдгара Ледерера (3).

Fig. 3.4 Extract from the first article on thin-layer chromatography

The fundamental work of N. A. Izmailov and M. S. Schreiber was called the "Drop-chromatographic analysis method and its application in pharmacy," a title given to an article which was published in 1938 [22]. Even the title of the article makes it clear that the authors understood this was a new method (Figs. 3.5 and 3.6).

In the late 1970s and early 1980s, Hungarian researchers (E. Tyihak et al.) described TLC with a closed sorption layer. This version of the method made it possible to greatly increase the speed of separation and improve the efficiency of the process. However, in practical implementation terms, this version was not very simple. The usual version of TLC, but with a "temporarily closed" sorption layer (see, e.g. [23]) was used by Berezkin et al. in 1980–1990. The method appeared to be more accessible in terms of practicality. This made it possible to avoid evaporation of the solvent from the plate and optimized the separation process. The development of these studies was of the back of the development of a method of electro-osmotic TLC on plates with a closed layer of sorbent.

B.G. Belenkii et al., in Leningrad, developed a high-performance TLC and transferred gel filtration to TLC (Fig. 3.7).

3.4 Development of Gas Chromatography

The idea of gas chromatography was expressed by A. Martin and R. Synge in 1941 [24]. It is believed that its GLC variant was carried out experimentally in 1952 by A. James and A. Martin [25]. However, the gas adsorption version of gas chromatography was realized earlier, and repeatedly, e.g., in the USSR in the late 1940s by N. M. Turkeltaub [26–28] (using the frontal method, M. M. Dubinin isolated gases in 1935–1936 [29, 30]).

In 1943–1944 N. M. Turkeltaub began work on the development of a chromatographic method for the determination of micro concentrations of hydrocarbons

Fig. 3.5 Nikolai Arkad'evich Izmailov (June 22, 1907–October 2, 1961), co-author of the first article on thin-layer chromatography, at a celebration of the 150th anniversary of Kharkov University (1955). He is on the far left, wearing a hat. Photo previously published

Fig. 3.6 Maria Semenovna Schreiber (1904–1981) was a pharmacist and helped develop thin-layer chromatography. Photo previously published

Fig. 3.7 Boris Grigor'evich Belenkii (May 12, 1926–May 8, 2008) developed high-performance thin-layer chromatography and contributed to the liquid chromatography of polymer and bio-organic substances. Photo provided by B. G. Belenkii

in air. He set himself the task of creating a highly sensitive and precise procedure that could be used for analysis in the field using a compact device. For the stationary phase, coal was used, with air being used as a carrier gas. Turkeltaub studied the possibility of separating mixtures of methane, ethane, and propane, depending on their total concentration and mixture composition, adsorbent humidity, and temperature. The adsorption isotherms of methane and ethane on dry and wet coal were obtained, and the dependence of the shape of the output curves on the adsorbent grain and air flow velocity was studied. Turkeltaub found that optimal conditions for the analysis of a mixture of three gases were achieved with the use of activated carbon, of the KAD brand, with a moisture content of 12 wt%. Under these conditions, the first portion of the gas produced contained all methane, the next did not contain any hydrocarbons, and the latter contained all ethane (Fig. 3.8).

Later N. M. Turkeltaub, improving the methods of chromatographic analysis of mixtures of hydrocarbons, carried out a series of studies on the theory of gas chromatography. Thus, he showed that in the maximum of the chromatographic peak under steady-state conditions, adsorption equilibrium always takes place. On this basis, he developed a method for the determination of the thermal parameters of adsorption. Turkeltaub also carried out a lot of work on the introduction of chromatographic methods to various branches of science and technology.

Nusin Motelevich Turkeltaub *(1915–1965) was a doctor of chemical sciences and laureate of the State Prize of the USSR. From 1944 he headed a laboratory in the oil industry. In 1948, in order to develop the geochemical methods of exploration of oil and gas fields, he developed a chromatographic method for the separation of hydrocarbons, designed to determine their traces in air* (Fig. 3.9).

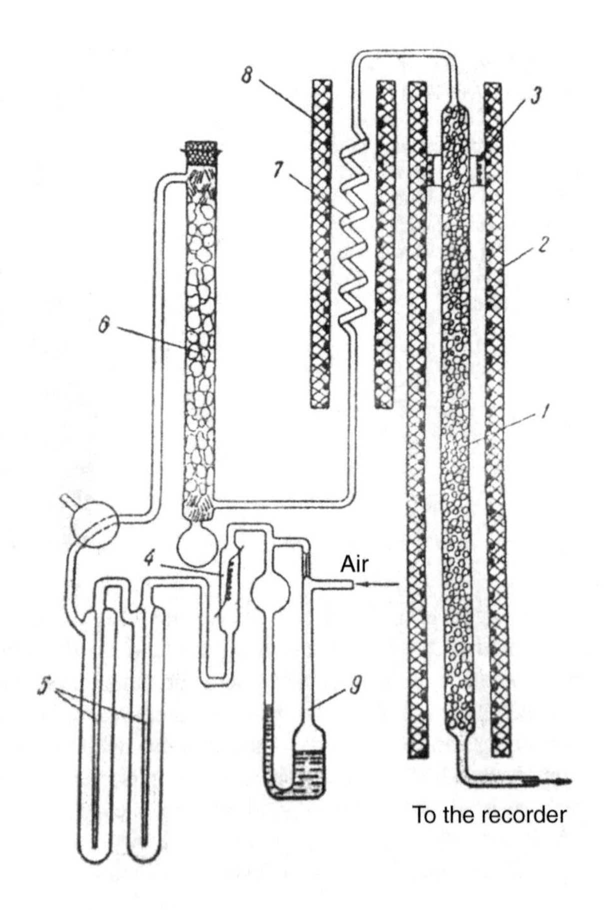

Fig. 3.8 Scheme of Turkeltaub's gas chromatograph. With permission of the journal *Zavodskaya Laboratoriya* (Plant Laboratory)

Fig. 3.9 Nusin Motelevich Turkeltaub (1915–February 2, 1965) was one of the first to develop gas chromatography in the late 1940s. Image provided by a former colleague of Turkeltaub

Much for gas chromatography was done by A. A. Zhukhovitskii, alongside N. M. Turkeltaub. Together with N. M. Turkeltaub he wrote the first domestic monograph on gas chromatography. Zhukhovitskii's work on the theory of the broadening of chromatographic zones and the separation of substances, and on the detection and new methods of chromatography, some of which have not been realized to date, far outstripped the development of chromatography and chromatographic instrumentation. Zhukhovitskii considered the thermal factor, which makes it possible to change adsorption properties of the sorbent over time and along the length of the column, to be of prime importance. Therefore, he attached special importance to the development of thermal methods. New variants of chromatographic analysis were developed—chromothermography, the thermodynamic method, and a method for multiple thermal desorption concentration. Owing to narrowing the chromatographic peaks, these methods make it possible to achieve high concentration factors, increase the efficiency of separation, and shorten the analysis time. Zhukhovitskii also proposed vacant, differential, and iteration chromatography, along with dosing of a pure carrier gas. A method of chromadistillation, based on multiple evaporation and condensation in a column with an inert filler or in a hollow tube with the isolation of pure components, was proposed. This method makes it possible to determine high-boiling compounds and microimpurities without prior concentration (Fig. 3.10).

***Alexander Abramovich Zhukhovitskii** (1908–1990) was a doctor of chemical sciences, a professor, and Honored Worker of Science and Technology. He graduated in 1930 from Novocherkassk Polytechnic Institute, and for 18 years worked at the L. Ya. Karpov Physicochemical Institute in Moscow. From 1948 he headed the Department of Physical Chemistry at the Moscow Institute of Steel and Alloys. He worked in the field of physical chemistry (physical chemistry of metallurgical processes, sorption processes, and surface tension) and gas chromatography. He carried out research on the theory of broadening of chromatographic zones and separation of substances as well as on detection and new methodologies of gas chromatography. He contributed to the organization of research on chromatography in a number of institutions that later became centers for the*

Fig. 3.10 Aleksandr Abramovich Zhukhovitskyii (September 5, 1908–December 19, 1990) developed of a number of gas chromatography variants and the method of chromadistillation, and co-authored the first monograph on gas chromatography in Russian. Photo previously published

development of gas chromatography. Among these are the All-Union Research Institute of Integrated Automation of the Oil and Gas Industry (VNIIKANeftegaz), the All-Union Scientific Research Geological and Petroleum Institute (VNIGNI), and the All-Union Scientific Institute of Nuclear Geology and Geophysics (VNIYAGG). On his initiative, the All-Union Research Institute of Chromatography (VNIIKhrom) was established. Together with his staff about 300 articles and 5 books were published; with more than 50 inventions to their names. He was chairman of the gas chromatography section of the Scientific Council of the USSR Academy of Sciences on chromatography. Among his students were more than 150 candidates and doctors of sciences. Zhukhovitskii was awarded the international M. S. Tswett Medal.

In 1936 A. V. Kiselev (Fig. 3.11) interpreted the "structural water" of silica gel as a hydroxyl coating, which formed the basis of the chemistry of surface compounds of silica. These studies served as impetus for the development of a new direction in the theory of surface phenomena, which was called surface chemistry.

Fig. 3.11 Andrei Vladimirovich Kiselev (November 28, 1908–July 17, 1984) was a specialist in molecular adsorption. He developed a classification of sorbents for gas chromatography. He also proposed new adsorbents. Photograph provided by the Chemistry Department of the Lomonosov Moscow State University

Kiselev and co-workers first used various structures for chromatography absorbents, like silica gels, silochromes, and the like (1962), as well as adsorbents based on carbon, like graphitized carbon black (1961), carbochromes (1974), etc. Chemical (1950) and geometric (1962) modification of adsorbents were used, making it possible to create the necessary surfaces. Adsorbents were suitable not only for analytical purposes, but also for physicochemical studies of a surface (surface area, heat of adsorption, and adsorption isotherms). Owing to numerous studies of adsorption and adsorption thermodynamics, A. V. Kiselev in 1967 created a well-known classification of adsorbents [31] (Fig. 3.12).

***Andrei Vladimirovich Kiselev** (1908–1984) was a physicochemist and professor at the Lomonosov Moscow State University. He was in charge of the Laboratory of Adsorption and Gas Chromatography. He was awarded the international M. S. Tswett Medal. Several his books are known* (Fig. 3.13).

In 1970–1990 V. G. Berezkin carried out theoretical and experimental studies on GLC, the results of which appeared to be important for the formation of the basis of the method [32, 33]. V. G. Berezkin's studies of the mechanism of the retention of sorbates in GLC were essential for understanding the process; they contributed to the theoretical interpretation of the separation process in GLC. A three-term equation proposed by Berezkin for the retention volume in GLC, taking into account the polyphase nature of real stationary phases, is used in modern chromatography and is known as the "Berezkin's equation." Berezkin investigated the influence of the nature of the carrier gas on the amount of sorbate retention in GLC

Fig. 3.12 The Chemistry Department of the Lomonosov Moscow State University. Picture provided by the author

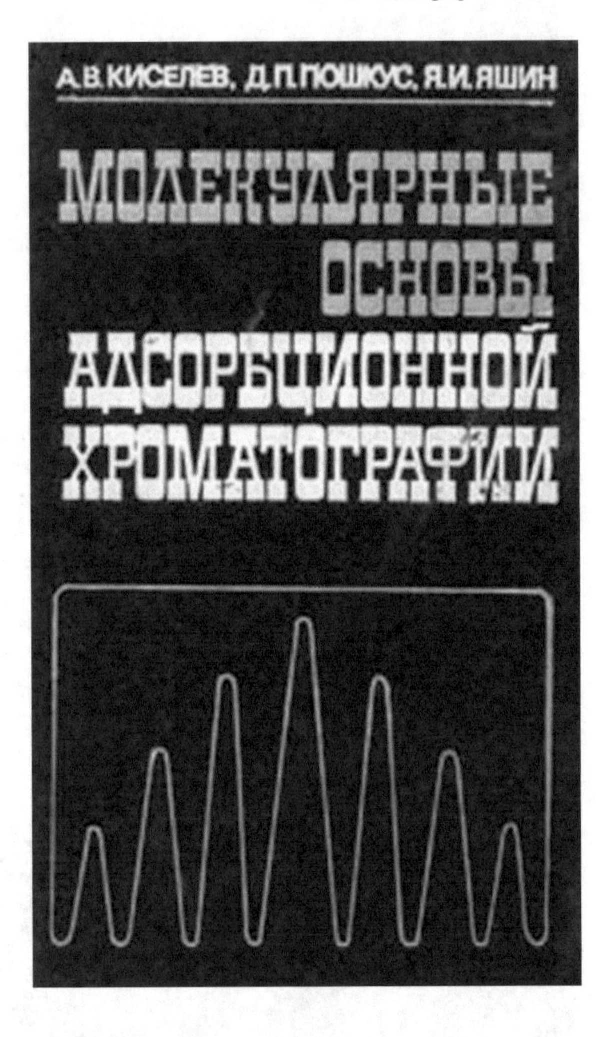

Fig. 3.13 One of A. V. Kiselev's books devoted to gas chromatography

[34]. It was shown that the opinion that the carrier gas does not affect the distribution processes and the selectivity of the separation of substances in the column was incorrect. To describe the dependence of the equilibrium chromatographic values on the average pressure of the carrier gas in the column, linear equations, which were in good agreement with experimental values up to pressures of 10 atm, were obtained. Berezkin also proposed new methods for the regulation of separation process in GLC. Under the guidance of Berezkin, a method was developed for studying phase transitions in polymers by GLC methods, which is widely used today. On his initiative, original quartz columns with an aluminum coating were developed. Berezkin was among the first chromatographers to develop reaction gas chromatography [35].

Viktor Grigor'evich Berezkin *was born April 18, 1931. He graduated from the chemistry department of Lomonosov Moscow State University (1954).*

He is a doctor of chemical sciences, a professor, Principle Researcher of the A. V. Topchiev Institute of Petrochemical Synthesis of the RAS, Honored Petrochemist of the USSR, laureate of the State Prize of the USSR, and Honored Worker of Science of Russia. His areas of scientific interest included: chromatographic methods of separation and concentration, theory of gas and planar chromatography, and analytical equipment. He coordinated more than 50 candidate of science theses and was the scientific consultant for three doctoral students. He authored more than 500 scientific articles and 20 books, many of which were translated in the United States, Holland, England, Germany, Hungary, and Poland, as well as 120 patents (Fig. 3.14).

Headspace analysis in gas chromatography was developed in 1960–1970 in the Leningrad University by B. V. Ioffe, A. G. Vitenberg, and B. V. Stolyarov [36]. They developed its theory and its numerous practical applications. Of great renown, and carrying international significance, were the monographs entitled *Gas Extraction in Chromatographic Analysis* (1982) and *Head-Space Analysis and Related Methods in Gas Chromatography* (Wiley, 1984). A. G. Vitenberg authored more than 150 publications on the establishment of regularities of gas extraction, along with having a series of patents. The activity of Vitenberg made it possible to convert the headspace analysis from a sample preparation variant to an independent method for studying condensed media containing volatile components. This work allowed to use the headspace method for measuring the solubility of volatile substances, for the theoretical description of the equilibrium model of gas extraction, to identify the characteristics of the corresponding physicochemical processes, for the development of headspace sources of gas mixtures for calibration and verification of analytical apparatus, and for gas chromatographic determination of volatile, sulfur-containing substances in industrial emissions and aqueous media.

Alexander Grigor'evich Vitenberg *(1935–2014) was a doctor of chemical sciences and a professor. He worked in the Gas Chromatography Laboratory of the Chemistry Department of St. Petersburg State University, created by B. A. Ioffe in 1968. From the early 1960s, A. G. Vitenberg, a graduate of the Leningrad*

Fig. 3.14 Viktor Grigor'evich Berezkin is one of the developers of reaction gas chromatography; he contributed to the theory of gas chromatography by demonstrating the role of the carrier gas. Photo provided by B. G. Berezkin

Chemical and Pharmaceutical Institute, worked at the University's Chemistry Department. In 1965, he defended his candidate of science thesis on the reactions of diazo compounds, in which the gas chromatography became one of the main methods of analysis, and in 1987 he defended his doctor of science thesis entitled "Gas extraction in headspace chromatographic analysis." He co-authored A Guide to Practical Works on Gas Chromatography (along with B. V. Stolyarov and I. M. Savinov), which ran to three editions (1973, 1978, and 1988). The supplemented and significantly expanded version of this manual, entitled Practical Gas and Liquid Chromatography, was published in 2002.

Ya. I. Yashin et al. [37] compiled a summary of the domestic achievements in gas chromatography (Table 3.1).

Table 3.1 Domestic achievements in the development of new methods of gas chromatography and the improvement of known methods

Creation of new methods and variants of chromatography or development of known methods	Author(s)	Year	Reference
Chromatothermography	A. A. Zhukhovitskii, N. M. Turkeltaub	1951	[38]
Heat dynamic method	A. A. Zhukhovitskii	1953	[39]
Volumetric chromatographic method	D. A. Vyakhirev	1953	[40]
Vacant chromatography	A. A. Zhukhovitskii	1962	[41]
Stepwise chromatography	A. A. Zhukhovitskii, N. M. Turkeltaub	1962	[42]
Capillary adsorption chromatography	V. I. Kalmanovskii, S. P. Zhdanov, A. V. Kiselev, M. M. Fiks	1962	[43]
Iteration chromatography	A. A. Zhukhovitskii	1963	[44]
Differential chromatography	A. A. Zhukhovitskii	1966	[45]
Chromatography without carrier gas	A. A. Zhukhovitskii	1972	[46]
Chromadistillation	A. A. Zhukhovitskii	1978	[47]
Chromatoscopy	A. V. Kiselev	1978	[48]
Circulation chromatography	V. P. Chizhkov, N. V. Sterkhov	1991	[49]
Development of gas adsorption chromatography	A. V. Kiselev, Ya. I. Yashin, K. D. Shcherbakova, Yu. S. Nikitin et al.	1959–1984	[50]
Development of reaction chromatography	V. G. Berezkin, V. R. Alishev, K. V. Alekseeva	1965–1970s	[35, 51]
Development of gas chromatography–mass spectrometry for the determination of impurities	I. A. Revel'skii, Yu. S. Yashin, A. I. Revel'skii	1970–2016	[52]

(continued)

Table 3.1 (continued)

Creation of new methods and variants of chromatography or development of known methods	Author(s)	Year	Reference
Development of gas extraction (analysis of equilibrium vapor)	B. V. Ioffe, A. G. Vitenberg, B. V. Stolyarov	1960–1980s	[53]
Development of gas chromatography on liquid crystals	M. S. Vigdergauz, L. A. Onuchak	1980s	[54]
New methods for calculating gas chromatographic retention indices. Relationship between retention indices and the structure of molecules. Identification by retention indices	I. G. Zenkevich	1990–2016	[55]
Classification of stationary liquid phases by thermodynamic parameters	R. V. Golovnya	1970–1980s	[56]

With permission from [37]

Several detectors were proposed, mainly for gas chromatography, in part for HPLC. They inlcude an uncontaminated electron capture detector (E. B. Shlendel' and V. A. Ioonson), a highly selective flame-photometric detector with a wide linear range (V. A. Ioonson), a system for microcoulometric detection of F, Cl, Br, I, S, P, and N-containing compounds (I. A. Revel'skii, Kh. E. Aavin, and P. Keres), a highly stable density detector for industrial chromatographs (V. N. Lipavskii), a diaphragm density detector (V. P. Guglya and A. A. Zhukhovitskii), a high-temperature (up to 850 °C) diaphragm detector using density (Ya. V. Mulyarskii and I. A. Revel'skii), and an amperometric detector for HPLC (Ya. I. Yashin).

One can add a few words about sorbents for gas chromatography. As already mentioned, A. V. Kiselev, a professor at Moscow University, created in 1962–1967 a classification of sorbents based on the mechanism of their action. Yu. S. Nikitin et al. proposed and studied silochromes and graphitized carbon black as chromatographic adsorbents; A. V. Kiselev and Ya. I. Yashin proposed surface-porous sorbents.

3.5 Liquid–Gas Chromatography and Chromatomembrane Methods

In 1982 L. N. Moskvin, A. I. Gorshkov, and M. F. Gumerov described liquid–gas chromatography [57–59], the latter variant of possible combinations of mobile and stationary chromatographic phases. The stationary gas phase is fixed in this case in narrow capillaries of the carrier, from which it is not replaced by a liquid that does not wet the carrier. The gases dissolved in the liquid are distributed between the mobile liquid and the gas that is in the capillaries. In 1984 Giddings and Myers published an article [60] on this method with its theoretical justification and an assessment of its possibilities.

Later, the authors of this method noted [61] that its analytical potential was limited. However, the discovery and study of this variant of chromatography paved the way for a more important process—the chromatomembrane process and various methodologies based on it.

The idea of a chromatomembrane process and associated methodologies was expressed and implemented by L. N. Moskvin in 1990. The first publications describing the idea, that is, the essence and conditions for the implementation of chromatomembrane mass exchange processes (CMMP), in liquid–liquid and liquid–gas systems, appeared in 1994–1996 [62–67]. Prior to these publications, several technical solutions based on chromatomembrane principles of mass transfer were patented: methods for determining volatile substances dissolved in liquids based on CMMP in a liquid–gas system, as well as methods for carrying out mass transfer and a device for its implementation, based on the principles of CMMP in liquid–gas and liquid–liquid systems. In the first publications on CMMP, the main areas of its application in analytical chemistry were outlined: the continuous and discrete release of determined substances from the flow of the analyzed polar liquid phase into the flow of the nonpolar liquid phase (liquid extraction) or gas phase flow (gas extraction), and also from the flow of the analyzed gas into the liquid phase (liquid absorption). The prospects of chromatomembrane concentration of analytes in combination with flow-injection analysis are experimentally substantiated in [68] (Fig. 3.15).

Leonid Nikolaevich Moskvin *was born on December 12, 1936. He graduated from the Chemistry Department of the Leningrad University (1959). He is a doctor of chemical sciences and a professor. He was in charge of the Division of Analytical Chemistry of St. Petersburg State University. L. N. Moskvin is the chairman of the St. Petersburg Branch of the Scientific Council of the Russian Academy of Sciences on Analytical Chemistry. He is an Honored Scientist of*

Fig. 3.15 Leonid Nikolaevich Moskvin developed the chromatomembrane method and contributed to liquid–gas chromatography. He is an expert in separation and concentration methods. Photo provided by L. N. Moskvin

Russia, honorary professor of St. Petersburg State University, and laureate of the V. G. Khlopin Prize from the Academy of Sciences and the RF Government Prize. He was awarded three orders.

His areas of scientific interest included methods of separation and preconcentration, chromatographic and flow analysis methods, radio analytical methods, and chemical and radiochemical technologies in nuclear power engineering. He carried out work in the field of extraction chromatography. He developed the analytical direction of continuous two-dimensional chromatography and carried out the first work in chromatography on block (monolithic) supports and sorbents. L. N. Moskvin authored *Liquid–Gas Chromatography*. He was one of the first to begin work on dialysis and electrodialysis through liquid extraction membranes impregnated on inert carriers, he developed methods of electroosmotic filtration for the deionization of water and preconcentration of electrically charged impurities from aqueous solutions, and developed countercurrent electrophoretic separation of ions with different isotopic composition. He formulated the idea of the chromatomembrane mass transfer process and chromatomembrane methods of liquid and gas extraction, as well as liquid absorption, based on its principles. He developed a scheme for rapid radiochemical analysis, embodied in the form of methods of rapid chromatographic radiochemical analysis and rapid membrane-sorption radiochemical analysis, which made it possible to automate radiochemical monitoring in nuclear power engineering.

The chromatomembrane method is based on the capillary effects in hydrophobic porous media. Mass exchange between the flows of immiscible liquids or liquid and gas is realized in a porous medium from a hydrophobic material with open pores. The independent motion of the flows of the two phases is due to the fact that the porous medium has two types of pores (macropores and micropores), which differ significantly in size. Macropores are chosen such that the capillary pressure in them is negligible and does not prevent the passage of the polar liquid phase. Micropores, on the contrary, are so small that capillary pressure prevents the polar liquid phase from penetrating them. At the same time, they should ensure sufficient permeability of the porous medium for the flow of a gas or a nonpolar liquid (Fig. 3.16).

The chromatomembrane process can be implemented in two modes: (1) continuous mode—when flows of two phases pass through the cell simultaneously; (2) discrete mode—when flows of two phases are sequentially passed through a chromatomembrane cell (CMC) with an overlapping of the channels at the input and output from the cell of that phase that is stationary at the moment. Microporous polytetrafluoroethylene (PTFE) membranes are used to input/output a flow of a nonpolar liquid phase or gas from a CMC. This material has maximal contact angles of wetting with aqueous solutions. Macropores vary in the range of 0.1–1.0 mm, depending on the desired permeability of the cell for the aqueous solution. The micropore sizes are 0.1–0.5 μm. For example, flow-injection schemes for determining SO_2, NO_2, and ammonia in air with photometric (in the determination of SO_2 and NO_2) and iodometric (in the ammonia emission) detection of ions formed in the absorption solution were developed on the principles of the chromatomembrane mass transfer process.

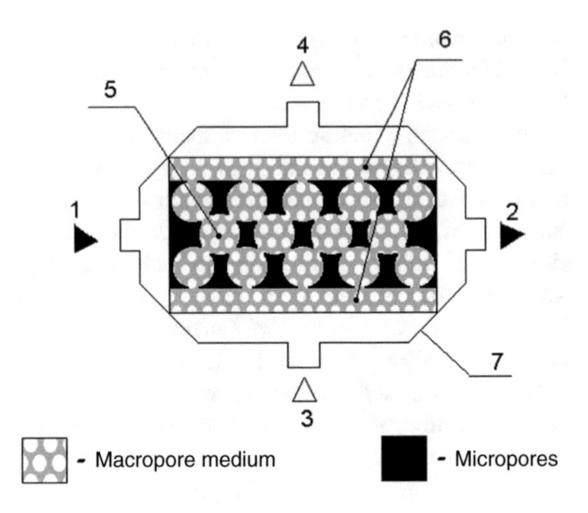

Fig. 3.16 Scheme for the chromatomembrane method. With permission of the Russian Academy of Sciences

Chromatomembrane headspace analysis has been realized and its advantages are shown in comparison with the traditionally used schemes. The efficiency of liquid chromatomembrane extraction is illustrated by considering examples of the photometric determination of microconcentrations of nitrite ions, phenol, and anionic surfactants, as well as the luminescent determination of petroleum products and phenols in natural water.

3.6 Enantioselective Ligand-Exchange Chromatography

Ligand-exchange chromatography was first described in 1961 by F. Helfferich in the journal *Nature* [69]. The possibility of separating optical isomers by this method was shown by V. A. Davankov [70–72], who became the founder of chiral (enantioselective) ligand-exchange chromatography (Fig. 3.17).

The sorbent matrix was chloromethylated polystyrene, synthesized by V. A. Davankov and S. V. Rogozhin in 1966. Amino acid L-proline was grafted onto this matrix due to the complexing amino acid group, and the obtained sorbent was treated with an ammoniacal solution of copper sulfate. The result was an optically active sorbent having the structure shown in Fig. 3.18. Using this method it became possible to completely separate the D and L-isomers of proline.

A story of V. A. Davankov's first work in this field is interesting. The following quotation is taken from an interview between him and his colleagues from Ufa [73]. In the interview, V. A. Davankov talks about entering the graduate course of academician Korshak at the Institute of Organoelemental Compounds of the USSR Academy of Sciences (INEOS).

"I immediately did not like the topic proposed by Korshak. Almost two years of postgraduate studies passed, but nothing sensible from the attempts to polymerize diketopiperazines did not work, even the second topic was proposed, but I decided to leave. Then they let me do 'whatever you want'.

... I totally devoted myself to the idea of separating the optical isomers of amino acids, especially as INEOS was actively developing methods for the synthesis of amino acids, and synthesis always results in a racemate, i.e., a mixture of two isomers. The country urgently needed L-amino acids to overcome the protein deficiency of feed (later the problem was solved by microbiological synthesis of L-amino acids).

I quickly obtained the pellets of styrene-divinylbenzene copolymers, introduced active chloromethyl groups into them, and replaced the chlorine atoms first with iodine, and then with nitrogen of optically active natural amino acids. On the columns with such optically active ion-exchange sorbents, I began to divide the racemates of amino acids, but was disappointed by the relatively low efficiency of separation. Turning to the literature, I was surprised to find about three hundred papers on earlier attempts to chromatographic separation of racemates, mostly unsuccessful or only 'promising'. The results I received were more than enough to defend the candidate of sciences thesis in 1966, but they had no practical perspective. I understood that we needed a fundamentally new approach that would

Fig. 3.17 Vadim Aleksandrovich Davankov developed ligand-exchange enantioselective chromatography and created universal sorbents based on super-cross-linked polystyrene. Photograph provided by V. A. Davankov

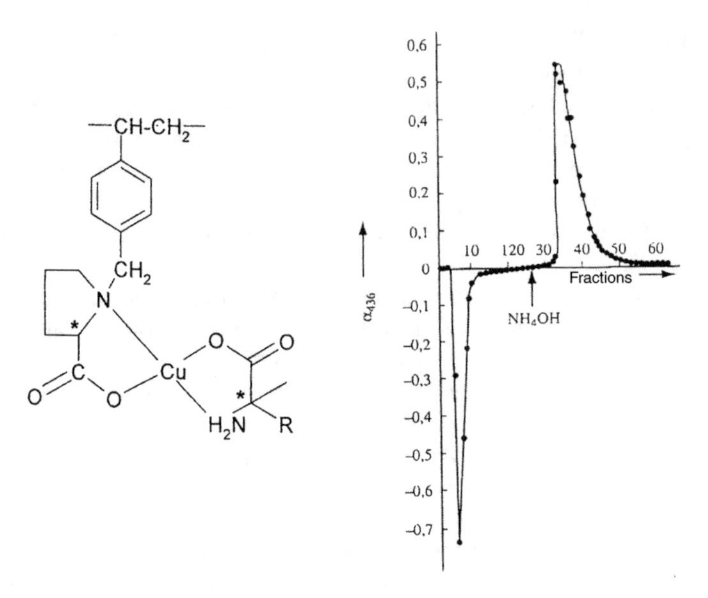

Fig. 3.18 Fragment of the first enantioselective sorbent and the first Davankov chiral chromatogram. Figure published previously

ensure a much closer, three-point contact of the separated isomers with the chiral selector, i.e., with optically active amino acid groups fixed on polystyrene.

My desperate search led me finally to ligand exchange chromatography, which turned out to be such a new approach to the separation of isomers, and I was the first to quantitatively separate the racemates of amino acids into their constituent isomers. The novelty was that I introduced complexing metal ions into the chromatographic system. Metal ions, primarily divalent copper, formed complexes simultaneously with the chiral groups of the sorbent and with the separated isomers of the amino acid. Embedded in the coordination sphere of the metal, both components came into close contact with each other and therefore clearly recognized the spatial structure of the partner. The sorbent with L-proline fixed groups predominantly bound to the complex only the D-isomers of the amino acid, whereas the L-isomers were easily leached from the column with water. The L-ligand bound to the complex was required to be displaced with ammonia. Therefore, the method is named ligand exchange chromatography. The enantioselectivity of such a chromatographic complexation process proved to be unprecedentedly high."

Vadim Aleksandrovich Davankov *was born on November 20, 1937. He graduated from the Technical University of Dresden (1962). He is a doctor of chemical sciences, a professor, head of the Laboratory of Stereochemistry of Sorption Processes at INEOS of RAS, and laureate of the State Prize of the Russian Federation. He was a titular member of IUPAC. He is a member of the Scientific Council of the Russian Academy of Sciences on High-Molecular Compounds. He was awarded the Martin Medal* (Fig. 3.19).

Fig. 3.19 Medal named after A. Martin, which was awarded to V. A. Davankov. Picture provided by V. A. Davankov

His research interests included polymer synthesis and chromatography. He proposed a new principle for the separation of enantiomers—ligand-exchange chromatography on chiral complex-forming sorbents. He proposed the principle of synthesis of super-cross-linked styrene polymers, on the basis of which a series of neutral polymeric sorbents were created. These sorbents possessed a unique high sorption capacity in relation to organic substances in water or air. They found wide application in the preconcentration of microimpurities, as well as in large-scale sorption processes in the food, chemical, and medical industries. V. A. Davankov authored more than 300 scientific works, including several books.

The conception and development of enantioselective ligand exchange chromatography was repeatedly considered in V. A. Davankov's reviews [74–76].

3.7 Critical Chromatography of Polymers

In the early 1980s, researchers at the Institute of Chemical Physics of the USSR Academy of Sciences, A. V. Gorshkov and V. V. Evreinov [77], developed a chromatographic method for the analysis of polymers—they called the method critical chromatography. The method made it possible to achieve significant progress in evaluating the molecular inhomogeneity of macromolecules and solving other problems linked to studying polymers. The method is based on the application of the theory of critical phenomena.

The main problem arising when studying the structure of macromolecules and determining "defectiveness" is the polydispersity of the samples. The presence of a distribution of molecular masses results in the fact that chromatograms of the macromolecules with different numbers of terminal groups and different masses overlap, and the separation of such molecules appears to be impossible. To see the difference in the chemical structure of polymers due to the presence of a small number of defects, it is necessary to "switch off" the interaction of the main monomers of the chain with the surface, and the associated separation dependence on the molecular weight distribution (MWD). The key to the solution of the problem is given by the phase character of the adsorption of polymers, which is a consequence of the boundness of the monomers in the chain and the collective character of their adsorption interaction. Owing to this, there are conditions, in which the MWD somewhat "disappears," becoming chromatographically invisible. These conditions, called critical conditions, are realized at the boundary of the adsorption and exclusion separation modes.

The term "critical" used for such a mode is based on the idea of the adsorption of macromolecules as a phase transition. At the critical point, the entropy losses are exactly compensated by the energy of attraction, and for a homogeneous chain this compensation takes place simultaneously for macromolecules of any length or molecular weight. This scale invariance, inherent in systems at the critical point, creates optimal conditions for studying the defectiveness of macromolecules. One can say that at the critical point only "visible" fragments, differing in the chemical structure from their main monomers, make a contribution to the separation—the pointwise of the functional group types, the spatial ones in the form of blocks, and also the way of their connection into the chain.

In fact, here we are talking about the possibility of separation at "the phase transition point". It turns out, however, that for polymers which are not too long, the phase transition point is blurred into a relatively wide region. Thus, the "nonideality" of the separation system turns out to be insignificant, and the critical mode may be realized on standard analytical systems. Since the energy of adsorption in HPLC depends mainly on the composition of the solvent, its smooth variation makes it possible to find a critical point in terms of its composition, in which the separation dependence on the molecular weight disappears. To fine-tune the system, temperature variation is additionally used.

A consequence of the phase nature of the adsorption transition is also the fact that the laws of separation of any macromolecules in the critical mode turn out to be universal. Since the losses of binding entropy, like the adsorption energy of a monomer, are determined by its chemical structure, the critical adsorption points for various polymers are different. However, if we "combine" the critical points for different polymers and different polymer–adsorbent–solvent systems in certain "dimensionless" coordinates, then the separation laws for various systems turn out to be similar. Such universality makes it possible to consider a single mechanism of separation in critical chromatography.

Adsorption chromatography of polymers is possible only near the critical point of adsorption. Even in the gradient version, long macromolecules remain adsorbed (immobile) until the solvent composition approaches the critical level. This allows the efficient separation of polymer mixtures according to the differences between their critical adsorption points.

This method has received theoretical and experimental justification and is used to solve various problems in the chemistry of high–molecular weight compounds, including problems of distribution by type of functionality, determination of composition and structure of block copolymers, separation of macromolecules by topology, the study of polymerization and degradation processes, separation of polymer mixtures, and the study of reactions involving macromolecules. This method has made it possible to study the structure of a chain of macromolecules and optimize the processes used to obtain polymers of a given structure. It is used not only in laboratory practice, but also in industry to control the quality of polymers and to identify problematic areas in synthesis technology which results in the appearance of defectiveness. Manufacturers of polymers, especially polymers with special properties, to some extent use this method, or a variant of it, in their research centers to solve technological problems and refine synthesis processes.

3.8 Polycapillary and Monolithic Chromatographic Columns

The desire to accelerate the gas chromatographic separation led, over time, to the creation of capillary columns by M. Golay. They rather quickly entered into practice, despite a number of limitations. One such limitation was that the gas chromatographs available at that time could not provide a significant increase in speed due to a discrepancy between the characteristics of the capillary columns and other components of the device – the input device, the detector, and the recording system [78]. Another limitation was that when using capillary columns with diameters of 10–50 μm, only a very small sample could be introduced into the column. This resulted in a narrowing of the range of determined concentrations (Fig. 3.20).

Fig. 3.20 One of the creators of polycapillary columns for chromatography, V. N. Sidel'nikov, with the author of this book at the fifth conference of "Analytics of Siberia and the Far East." Photo from the author`s collection

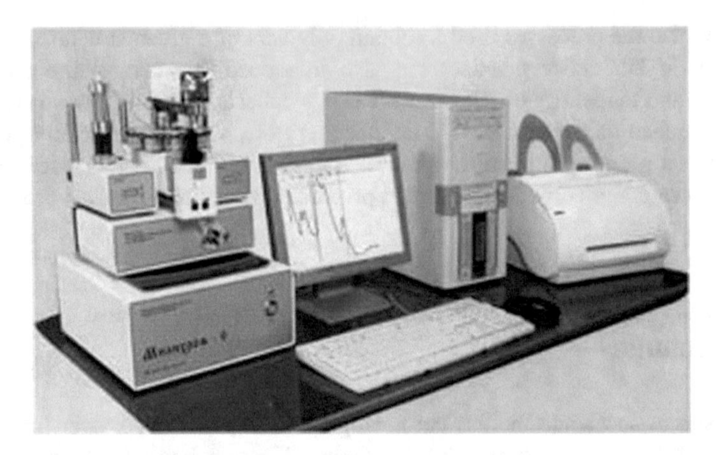

Fig. 3.21 "Milichrom-6 Leader" liquid microcolumn chromatograph. Photo taken from an advertising brochure

To overcome the second limitation, an idea was proposed to create columns containing a large number of small-diameter capillaries [79, 80]. Some of these publications even outlined ways to produce such columns. However, the realization of the idea turned out to be possible only after the development of sufficient technology to obtain polycapillary columns. This technology was provided by a group of specialists in Novosibirsk under the leadership of V. V. Malakhov [81]. An article by the Novosibirsk scientists in 1993 [82] was the first article on polycapillary chromatography. A second publication appeared (this time not in a scientific journal) in 1996 [83] (Fig. 3.21).

3.9 Countercurrent Chromatography in Inorganic Analysis

This method, called countercurrent chromatography in English-speaking literature, is called liquid chromatography with a free stationary phase, in Russian literature. It was developed primarily by Ito [84–86] for the separation of various kinds of bio-objects, as well receiving developmental input by Conway and co-workers [87–89]. To implement this method, planetary centrifuges of special design are used, the main element of which is a rotating, spiral column. This setup is comprised of a capillary, spirally wound on a cylindrical core, rotating about its axis and simultaneously revolving around the central axis of the device. Under the action of centrifugal forces, arising during planetary motion, one of the phases of the two-phase fluid system is retained in the column during the continuous pumping of the other phase. This method successfully combines the advantages of chromatography and dynamic multi-step extraction.

Fig. 3.22 Boris Yakovlevich Spivakov is a known specialist in methods of separation and preconcentration; he applied countercurrent chromatography to inorganic analysis. Photo provided by B. Ya. Spivakov

This method was applied by B. Ya. Spivakov et al. [90, 91] to separate inorganic compounds, primarily chemical elements, and received broad development in this direction. The method makes it possible to automate the multi-stage liquid–liquid distribution process and is used to obtain high-purity reagents as well as being used to analyze various objects [92, 93] (Fig. 3.22).

Boris Yakovlevich Spivakov *was born on May 19, 1941. He is a corresponding member of the Russian Academy of Sciences, a professor, and he was in charge of the Concentration Laboratory at the V. I. Vernadskii Institute of Geochemistry and Analytical Chemistry of the Russian Academy of Sciences. He is deputy chairman of the Scientific Council of the Russian Academy of Sciences on Analytical Chemistry. He was awarded the L. A. Chugaev and V. G. Khlopin Prizes of the Russian Academy of Sciences.*

He developed a theory of exchange extraction of chelates, contributed to the description of liquid extraction from the position of coordination chemistry, and proposed organotin compounds as extractants. He used extraction in two-phase aqueous systems and countercurrent chromatography in inorganic analysis. He developed methods for speciation analysis and the analysis of microparticles.

3.10 Other Work on Chromatography

V. I. Kalmanovsky and Ya. I. Yashin et al. developed many types of gas and liquid chromatographs and detectors, including the first portable ion chromatograph and a high-performance amperometric detector for HPLC. Ya. I. Yashin is a popularizer of chromatography and a historiographer of its Russian branch; he has pointed to a number of promising areas of development using chromatographic methods, which have given impressive results (Fig. 3.23).

Yakov Ivanovich Yashin *was born on July 30, 1936. For a long time he worked at Khimavtomatika Co., first in Dzerzhinsk, then in Moscow. He directed the company toward chromatography. In recent years he works for the "Interlab" company. He is a doctor of chemical sciences, a professor, and a member of the*

Fig. 3.23 Yakov Ivanovich Yashin developed a large number of chromatographs and is a specialist in gas and liquid chromatography; he proposed an effective amperometric detector for high-performance liquid chromatography. Photo provided by Ya. I. Yashin

editorial board of the Journal of Analytical Chemistry, as well as a member of the Bureau of the Scientific Council of the Russian Academy of Sciences on Analytical Chemistry.

A large volume of work on ion chromatography (Yu. A. Zolotov, O. A. Shpigun, M. M. Senyavin, A. M. Dolgonosov et al. were awarded the State Prize in 1991) has been performed. As early as 1990 the book *Ion Chromatography in the Analysis of Waters* was published in Russian and English. This method is widely used, for example, for water monitoring in nuclear power engineering and thermal power plants (Fig. 3.24).

Oleg Alekseevich Shpigun *was born on November 16, 1946. He is a corresponding member of the Russian Academy of Sciences, a professor of the Chemistry Department of Lomonosov Moscow State University, deputy head of the Division of Analytical Chemistry, deputy chairman of the Scientific Council on Analytical Chemistry of the Russian Academy of Sciences, and chairman of the Commission*

Fig. 3.24 Oleg Alekseevich Shpigun made a significant contribution to the development of ion chromatography and high-performance liquid chromatography, as well developing methods for determining the components of rocket fuels, bio-organic substances, etc. Photo provided by O. A. Shpigun

on Chromatography of the Council. He was laureate of the State Prize. He proposed using solutions of amino acids as eluents for ion chromatography. Together with his staff, he synthesized new effective sorbents for this method and for HPLC as a whole. He also helped develop chromatographic and chromatography–mass-spectrometric methods for the determination of the components of rocket fuels and a number of bioactive substances. O. A. Shpigun supervised the Analytical Center of the Department.

In the 1950s–1980s the development of chromatography in the USSR was largely facilitated by the works of K. V. Chmutov, K. I. Sakodynskii, O. G. Larionov, M. S. Vigdergauz, V. D. Vyakhirev, and many others. During this time, K. I. Sakodynskii showed great determination to perpetuate the memory of M. S. Tswett.

References

1. Rudenko, B.A. (ed.): 100 Years of Chromatography, p. 739. Nauka, Moscow (2003) (in Russian)
2. Tswett M.S.: Works and Proceedings of the Sessions of Society, Naturalists at the University of Warsaw, Department of biology, year 14, 1 (1903) (in Russian)
3. Tswett, M.S.: Veg. Dtsch. Bot. Geselsch. **24**, 316 (1906)
4. Tswett, M.S.: Ber. Dtsch. Bot. Geselsch. **24**, 384 (1906)
5. Senchenkova, E.M.: The Birth of the Idea and the Method of Adsorption Chromatography, 228 pp. Nauka, Moscow (1991) (in Russian)
6. Senchenkova, E.M., Tswett, M.S.: The Creator of Chromatography, 440 pp. Yanus-K, Moscow (1997) (in Russian)
7. Senchenkova, E.M.: Mikhail Semenovich Tswett, 307 pp. Nauka, Moscow (1973) (in Russian)
8. Ettre, L.S., Zlatkis, A. (eds.): 75 Years of Chromatography—A Historical Dialogue, 502 pp. Elsevier, Amsterdam (1979)
9. Ettre, L.S., Zlatkis, A. (eds.): 75 Years of Chromatography—A Historical Dialogue, pp. 53, 502. Elsevier, Amsterdam (1979)
10. Ettre, L.S., Sakodynskii, K.I.: Chromatographia **35**(223), 329 (1993)
11. Tswett, M.S.: Selected Works, Zolotov, Yu.A. (ed.), Senchenkova, E.M. (Coll.), 679 pp. Nauka, Moscow (2013) (in Russian)
12. Stahl, Y. (ed.): Dunnschichtchromatographie, 534 pp. Springer, Berlin (1962)
13. Schreiber, M.S.: The beginnings of thin layer chromatography. J. Chromatogr. **73**, 367 (1972)
14. Schreiber, M.S.: Discovery of chromatography in a thin layer. In: Chmutov, K.V., Sakodynskii, K.I. (eds.) Success of Chromatography. To the 100th Anniversary of the Founder of the Chromatography M.S. Tswett, p. 31. Nauka, Moscow (1972). (in Russian)
15. Izmaylov, N.A., Schreiber: Discovery of Thin-layer Chromatography, M.S., Berezkin, V.G. (ed.), 128 pp. GEOS, Moscow (2007) (in Russian)
16. Ettre, L.S., Zlatkis, A. (eds.): 75 Years of Chromatography—A Historical Dialogue, pp. 53, 413, 502. Elsevier, Amsterdam (1979)
17. Hamilton, R.J., Hamilton, Sh., Kealey, D. (ed.): Thin Layer Chromatography, 129 pp. Wiley, Chiechester (1987)
18. Guiess, F.: Basics of Thin-layer Chromatography, 405, 348 pp. Scientific Council on chromatography, Moscow (1999) (in Russian)

19. Guiess, F.: Fundamentals of Thin Layer Chromatography (Planar Chromatography), 482 pp. Huethig, Heidelberg (1987)
20. Keaus, L., Koch, A., Hoffstetter-Kuhn, S.: Dunnschichtchromatographie, 205 pp. Berlin, Springer (1996)
21. Ettre, L.S., Kalasz, H.: LC/GC, 19, 712
22. Izmailov, N.A., Schreiber, M.S.: Pharmaciya, p. 1 (1938)
23. Berezkin, V.G., Korshikina, E.V.: Zh. Anal. Khim. **61**, 1074 (2006)
24. Martin, A.J.P., Synge, R.L.M.: Biochem. J. **35**, 1358 (1941)
25. James, A.T., Martin, A.J.P.: Biochem. J. **50**, 679 (1952)
26. Turkel'taub, N.M.: Zavodsk. Lab. **15**, 653 (1949)
27. Turkel'taub, N.M.: Zh. Anal Khim. **5**, 200 (1950)
28. Zhukhovitskii, A.A., Zolotareva, O.V., Sokolov, V.A., Turkel'taub, N.M.: Dokl. AN SSSR **77**, 435 (1951)
29. Dubinin, M.M., Khrenova, M.: Zh. Prikl. Chem. **9**, 1204 (1936)
30. Dubinin, M.M., Yavich, S.: Zh. Prikl. Chem. **9**, 1191 (1936)
31. Kiselev, A.V., Poshkus, D.P., Yashin, Y.I.: Molecular Basis of Adsorption Chromatography, 272 pp. Khimiya, Moscow (1986) (in Russian)
32. Berezkin, V.G.: Chemical Methods in Gas Chromatography, 256 pp. Khimiya, Moscow (1980) (in Russian)
33. Berezkin, V.G.: Gas-Liquid-Solid-Phase Chromatography, 112 pp. Khimiya, Moscow (1986) (in Russian)
34. Berezkin V.G.: Ros. Khim. Zh. (Zh. Ross. Khim. ob.) **47**, 35 (2003)
35. Berezkin, V.G.: Analytical Reaction Gas Chromatography, 184 pp. Springer, Berlin (1966)
36. Berezkin, V.G., Ioffe, B.V.: Gas Extraction in Chromatographic Analysis. Headspace Analysis and Related Methods, 280 pp. Khimiya, Moscow (1982) (in Russian)
37. Yashin, Y.I., Yashin, E.Y., Yashin, A.Y.: Gas Chromatography, 528 pp. Trans Lith, Moscow (2009) (in Russian)
38. Zhukhovitskii, A.A.: Dokl. AN SSSR **77**, 435 (1951)
39. Zhukhovitskii, A.A., Turkel'taub, N.M., Georgievskaya, T.V.: Dokl. AN SSSR **92**, 987 (1953)
40. Vyakhirev, D.A., Komissarov, P.F.: Dokl. AN SSSR **129**, 138 (1953)
41. Zhukhovitskii, A.A., Turkel'taub, N.M.: Dokl. AN SSSR **143**, 646 (1962)
42. Zhukhovitskii, A.A., Turkel'taub, N.M.: Dokl. AN SSSR **144**, 829 (1962)
43. Zhdanov, S.P., Kalmanovskii, V.I., Kiselev, A.V., Fiks, M.M., Yashin, Y.I.: Zh. Fiz. Khim. **56**, 1118 (1962)
44. Zhukhovitskii, A.A., Turkel'taub, N.M.: Dokl. AN SSSR **150**, 113 (1963)
45. Zhukhovitskii, A.A.: Zavodsk. Lab. **32**, 402 (1966)
46. Zhukhovitskii, A.A.: Zh. Anal. Chim. **27**, 971 (1972)
47. Zhukhovitskii, A.A., Yanovskii, S.M., Shvartsman, V.P.: In the Collection "Chromatography", 2, p. 49. VINITI, Moscow (1978)
48. Kiselev, A.V.: Chromatographia **11**, 691 (1978)
49. Sterkhov, N.V., Chizhkov, V.P.: Zavodsk. Lab. **57**, 1 (1991)
50. Kiselev, A.V., Yashin, Y.I.: Gas Adsorption Chromatography, 254 pp. Plenum Press, New York (1969)
51. Berezkin, V.G., Gorshunov, O.L.: Usp. Khim. **34**, 1108 (1965)
52. Revelskii, I.A. In: Rudenko, B.A. (ed.) et al.: 100 years of Chromatography, p. 529. Nauka, Moscow (2003) (in Russian)
53. Vitenberg, A.G., Ioffe, B.V.: Gas Extraction in Chromatographic Analysis, 279 pp. Khimiya, Leningrad (1982) (in Russian)
54. Vigdergauz, M.S., Belyaev, N.F., Esin, M.S.: Fres. Z. Anal. Chem. **335**, 70 (1989)
55. Zenkevich, I.G. In: Rudenko, B.A. (ed.): 100 years of chromatography, p. 311. Nauka, Moscow (2003) (in Russian)
56. Golovnya, R.V., Misharina, T.A.: J. High. Resolut. Chromatogr. Commun. **33**, 51 (1980)
57. Moskvin, L.N., Gorshkov, A.I., Gumerov, M.F.: Dokl. AN SSSR **265**, 378 (1982)

58. Moskvin, L.N., Gorshkov, A.I., Gumerov, M.F.: Zh. Phis. Khim. **1983**, 57 (1979)
59. Gumerov, M.F., Rodinkov, O.V., Moskvin, L.N., Gorshkov, A.I.: Zh. Phis. Khim. **62**, 2249 (1988)
60. Giddings, J.C., Myers, M.N.: J. High Resolution. Chromatogr. **6**, 831 (1984)
61. Moskvin, L.N., Rodinkov, O.V.: Chromatomembrane Methods of Separation of Substances, St. Petersburg: St. Petersburg University, 2014, 216 pp. (in Russian)
62. Moskvin, L.N.: J. Chromatogr. A **669**, 81 (1994)
63. Moskvin L.N., Grigor'ev G.L.: Izv. Vyssh. Ucheb. Zaved. Tsvetn. Metally, 2026 (1994)
64. Moskvin, L.N., Rodinkov, O.V., Kartuzov, A.N.: Zh. Anal. Khim. **51**, 835 (1996)
65. Moskvin, L.N., Noktlaeva, D.N., Mikhailova, N.V.: Zh. Anal. Khim. **51**, 845 (1996)
66. Moskvin, L.N., Rodinkov, O.V.: J. Chromatogr. A **725**, 351 (1996)
67. Moskvin, L.N., Kartuzov, A.N., Tulupov, A.N. et al.: Patent of the Russian Federation No. 2023 488. Bul. Izobr. No. 22 (1994)
68. Moskvin, L.N., Simon, J.: Talanta **41**, 1765 (1994)
69. Helfferich, F.: Nature (London) **1961**, 189 (1001)
70. Davankov, V.A., Rogozhin, S.V.: Chromatographic method of cleavage of racemates of optically active compounds, Patent SSSR No. 308635, 1968 (in Russian)
71. Davankov, V.A., Rogozhin, S.V.: Dokl. AN SSSR **193**, 94 (1970)
72. Rogozhin, S.V., Davankov, V.A.: J. Chem. Soc. D; Chem. Commun. **490** (1971)
73. Khatymova, L.Z., Mazunov, V.A., Khatymov, R.V.: Istor. nauki i tekhn. No. 5, 73(spl 2) (2008)
74. Rogozhin, S.V., Davankov, V.A.: Usp. Khim. **37**, 1327 (1968)
75. Davankov, V.A., Kurganov, A.A., Rogozhin, S.V.: Usp. Khim. **43**, 1610 (1974)
76. Davankov, V.A., Kurganov, A.A., Bochkov, A.S. In: Giddings, J.C. et al (eds.): Advances in Chromatography, vol. 22, p. 7 (1983)
77. Gorshkov, A.V., Evreinov, V.V., Entelis, S.G.: Dokl. AN SSSR **272**, 632 (1983)
78. Sidel'nikov, V.N., Patrushev, Y.V.: Ros. Khim. Zhurn. (Zh. Ross. Khim. Ob.) **47**, 23 (2001)
79. Zhukhovitskii, A.A., Turkel'taub, N.M.: Gas Chromatography, 138 pp. Gostopmekhizdat, Moscow (1963) (in Russian)
80. Janik, A.: J. Chromatogr. Sci. **14**, 589 (1976)
81. Papent SSSR No. 986181, 1980
82. Malakhov, V.V.: Sidel'nikov V.N., Utkin V.A. Dokl. Akad. Nauk **329**, 749 (1993)
83. Cooke W.S.: Today's Chemist at Work. January 1996, 16
84. Ito, Y., Weinstein, M.A., et al.: Nature **212**, 985 (1966)
85. Ito, Y., Bowman, R.L.: Science **167**, 281 (1970)
86. Ito, Y.: J. Chromatogr. **301**, 377 (1984)
87. Conway, W.D.: Countercurrent Chromatography. Apparatus, Theory and Applications, 475 pp. VCH, New York (1996)
88. Ito, Y., Conway, W.D. (eds.): High-Speed Countercurrent Chromatography. Wiley, New York (1996)
89. Menet, J.-M., Thiebaut, D. (eds.): Countercurrent Chromatography (Chromatographic Science Series, 82), 171 pp. Marcel Dekker, New York (1999)
90. Pavlenko, I.V., Bashlov, V.L., Spivakov, B.Y., Zolotov, Y.A.: Zh. Anal. Khim. **44**, 827 (1989)
91. Bashlov, V.L., Pavlenko, I.V., Spivakov, B.Y., Zolotov, Y.A.: Zh. Anal. Khim., **44**, 1012 (1989)
92. Berthod, A., Maryutina, T., Spivakov, B., Shpigun, O., Sutherland, I.A.: Pure Appl. Chem. **81**, 355 (2009)
93. Spivakov, B.Y., Maryutina, T.A., Fedotov, P.S., Ignatova, S.N.: Different two-phase liquid systems for inorganic separation by counter current chromatography. In: Bond, A.H., Dietz, M.L., Rogers, R.D. (eds.) Metal-Ion Separation and Preconcentration, p. 333. Washington. (1999)

Chapter 4
Electrochemical Methods

Abstract There are at least three areas in which Russian electroanalysts have made a significant contribution: (1) ion-selective electrodes (theory, electronic tongue, etc.); (2) inverse voltammetry (analytical signal theory and as well as numerous other specific procedures); and (3) new electrodes (especially modified ones, for use in various electrochemical methods). Specialists in this field include B. P. Nikol'skii, Yu. G. Vlasov, A. G. Stromberg, Kh. Z. Brainina, and G. K. Budnikov. A precision coulometer has also been developed in Russia.

4.1 General Remarks

A number of the achievements of Russian scientists in the field of electroanalytical chemistry are covered in the book compiled by Scholz [1]. The results of many studies on voltammetry and coulometry are considered in reviews by Budnikova and Shirokova [2, 3]; a number of the important aspects of the history of this field of science are covered in the book by Compton et al. [4] as well as being included in [5]. F. Scholtz wrote separately about female Russian electroanalysts [6]. In these publications, you can find many technical details, as well as information about the researchers themselves (the latter especially applies to the publications of F. Scholz).

In Russia, there were powerful groups on electrochemistry-theoretical (A.N. Frumkin et al) and technical (N.A. Izgaryshev et al); this largely contributed to the research of Russian analysts in this direction.

Russian achievements in the field of electrochemical methods of analysis can be attributed to:

1. The development of theory, and the creation of new ion-selective electrodes (ISEs) and their ensembles, including B. P. Nikol'skii's ion-exchange theory of ISEs and the development of the electronic (artificial) tongue by Yu. G. Vlasov and co-workers.
2. The development of the theory of inverse voltammetry (IV) and the application of this method to solve a large number of practical problems (A. G. Stromberg, Kh. Z. Brainina, and many others).

Y. A. Zolotov, *Russian Contributions to Analytical Chemistry*,
https://doi.org/10.1007/978-3-319-98791-0_4

3. Successes in the development of some other electrochemical methods (polarography of organic compounds in non-aqueous media, electrochemical biosensors, and coulometry). One can specifically mention Yu. M. Kargin and G. K. Budnikov, and if you include colleagues from the former Soviet republics, then Ya. P. Stradyn, M. T. Koslovskii, O. A. Songina, and V. D. Bezuglyi.

In the 1930th–1970s a large number of studies have been carried out on classical polarography, encluding applied approaches with the introduction of a number of procedures into its broader practice [7, 8]. If at the beginning of this period the problems of metallurgy or the extraction and processing of mineral raw materials were being solved for the most part, then the interest was to a great extent shifted to objects of organic nature. The area of organic electrochemistry, initiated by A. N. Frumkin, was formed; in the 1960s the first conferences on organic electrochemistry took place. The results of theoretical research in this field were in demand from applied work, and gradually polarography and then voltammetry became widely used in relation to organic objects.

4.2 Ion-Selective Electrodes and the Electronic Tongue

In the mid 1930s, the Leningrad chemist and future academician Boris Petrovich Nikol'skii, who was in exile in Saratov, developed the so-called ion-exchange theory of ISEs [9, 10]. He derived an equation that is widely used today. Later, in Leningrad, when he returned from exile, he created a school of physicochemists, actively engaged in work on glass electrodes, mainly for measuring pH. For the development of this branch of research, and the development and organization of the production of such electrodes, the Leningrad group was awarded the USSR State Prize (Fig. 4.1).

Fig. 4.1 Boris Petrovich Nikol'skii (October 14, 1900–January 04, 1990) was probably the first to develop the theory of ion-selective electrodes. Photo previously published

Work on ISE was carried out (and successfully developed) not only at the Division of Physical Chemistry of the Leningrad University, which was headed by B. P. Nikol'skii, but also at the Department of Radiochemistry at the same university, headed by Professor Yu. G. Vlasov. Work was also carried out at the Saratov University (E. A. Materova, E. G. Kulapina, et al.) and the D. I. Mendeleev Russian Chemico-Technological University (O. M. Petrukhin et al.). Yu. G. Vlasov investigated the dependence of the parameters of solid-state chemical sensors on the physical and chemical properties of solid membranes (films and bulk samples). Regularities in the influence of various factors (structures, ionic and electronic conductivity, ionic conductivity type, and type and concentration of defects in membrane materials) on the mechanisms of functioning of solid-state chemical sensors were identified. The results of fundamental research in solid-state physicochemistry were used in the development of chemical sensors of new types (ion-selective field-effect transistors and chalcogenide glass sensors). Many procedures were developed and used for the determination of different ions.

These identified regularities made it possible to use weakly selective sensors in analyzers of liquid media. These sensors were given the name "electric tongue," as suggested by Vlasov and co-workers (including Italians). Sensor systems are used in combination with mathematical methods of pattern recognition (artificial neural networks, etc.) (Fig. 4.2).

The first multisensor system of "electronic tongue" type was a "taste sensor," proposed by Japanese scientists in the early 1990s [11, 12]. It included 8 potentiometric sensors with lipid membranes that had cross-sensitivity to various substances. Field transistors, potentiometric sensors with variable surface photopotential, were used as transducers. Vlasov and other researchers used ISEs of various types, including those based on chalcogenide glasses [13, 14]. The "electronic tongue" has been used to solve a number of environmental problems as well as for the classification and determination of food quality (Fig. 4.3).

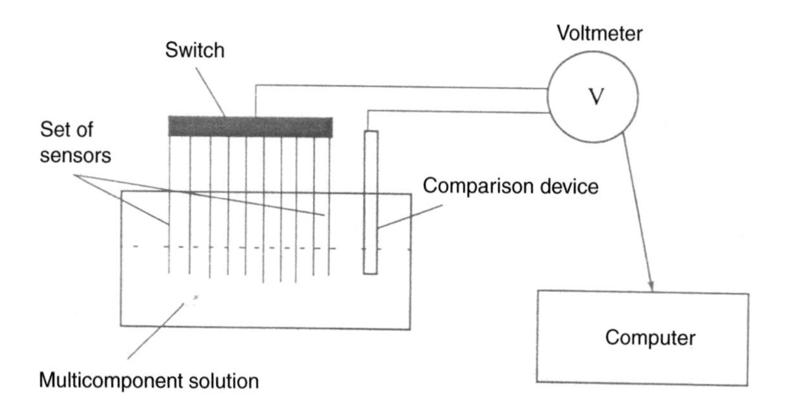

Fig. 4.2 Scheme of the electronic tongue. With permission of the Russian Academy of Sciences

Fig. 4.3 Yurii Georgievich Vlasov (June 26, 1934–October 17, 2016) developed chalcogenide ion-selective electrodes, proposed the term "electronic tongue," and is one of the most cited Russian analysts. Photo provided by Yu. G. Vlasov

Yurii Georgievich Vlasov *(June 26, 1934–October 17, 2016) graduated from the Chemistry Department of the Leningrad University (1958). He was doctor of chemical sciences and a professor. He was in charge of the Division of Radiochemistry of the Chemistry Department of the St. Petersburg University. He was a member of the editorial boards of: Journal of Analytical Chemistry, Journal of Applied Chemistry, and Radiochemistry. He worked at the International Union of Pure and Applied Chemistry. Professor Vlasov was a member of the permanent organizing committee of "Eurosensor" conferences, Honored Worker of Science of the Russian Federation, Honored Worker of Higher Professional Education of the Russian Federation, and honorary professor of the St. Petersburg University. He authored more than 400 articles, including 4 chapters in monographs, and held 25 Russian and 2 US patents; he was one of the most cited Russian analysts.*

The work of Vlasov and co-workers (Legin and Rudnitskaya) on the electronic tongue, as well as the development of the electronic nose, conducted in many countries, stimulated rather wide Russian research in these and related areas, e.g., similar work was carried out by several groups in Voronezh [15, 16], in Ufa [17], and other places too.

4.3 Voltammetry (Especially Its Stripping Version)

In the 1930s, and subsequent years in the USSR, much attention was paid to the development and application of polarography. There were several working groups in the country—in Moscow, Leningrad, Kazan, Kishinev, Kiev, Tomsk, Sverdlovsk, Tyumen', Khar'kov, Alma-Ata, and other cities. Among the areas developed in this field were: the study and use of catalytic currents (S. G. Mairanovskii and V. F. Toropova); polarography in non-aqueous and mixed solutions (Ya. P. Stradyn' and Yu. M. Kargin); different emerging variants of polarography—pulse, square-wave, cyclic, etc.; determination of organic compounds; and the creation of new instruments (see below) and electrodes (O. A. Songina, O. L. Kabanova, Kh. Z. Brainina, et al.) [18, 19].

Much progress was made in the development of the general theory (T. A. Kryukova, R. Sh. Nigmatullin, M. R. Vyaselev et al.) and polarographic instrumentation (S. B. Tsfasman, B. S. Bruk, I. E. Bryksin, R. M. -F. Salikhdzhanova, G. I. Ginzburg et al.). Some work was pioneering in nature, e.g., fractional differentiation of a voltammogram and use of a stepwise polarizing voltage to eliminate capacitive current (Nigmatullin). Work on the non-aqueous polarography of organic compounds (Kargin) made it possible to look differently at the integral form of a multielectron wave, adopting the concept of included chemical reactions that occur simultaneously with the actual electron transfer. It turns out that these results are important for interpreting curves (the amperometric response) in the framework of quantitative analysis using voltammetry.

A new powerful wave of research has risen after the appearance of stripping voltammetry (SV). After the first work in this field at the M. V. Lomonosov Moscow State University (E. N. Vinogradova) and V. I. Vernadskii Institute of Geochemistry and Analytical Chemistry of the USSR Academy of Sciences (S. I. Sinyakova), extensive studies were carried out in Tomsk (A. G. Stromberg), as well as in Yekaterinburg (Kh. Z. Brainina) and many other research centers.

A notable achievement in the field of SV (mainly on mercury-film electrodes) is the development of the theory and the general methodology by A. G. Stromberg and co-workers (Fig. 4.4). A. G. Stromberg, who worked at the Tomsk Polytechnic Institute, was engaged in SV for more than 40 years, since 1960. He developed the parametric theory of the method, the theory of mixed potential in extremely dilute solutions, and used the method for studying intermetallic compounds of mercury and complex ions. A lot of attention was paid to the theory of analytical signals [4, 20]. In his laboratory he created a large number of analysis procedures, many of which were introduced by enterprises in the electronic and chemical industries, as well as those in non-ferrous metallurgy. A. G. Stromberg created a large scientific school; under his leadership, more than 100 candidate of sciences thesis and several doctoral dissertations were defended. It should be noted that studies using mercury and amalgam electrodes in SV are currently developing more slowly (Fig. 4.4).

Armin Genrikhovich Stromberg *(September 16, 1910–September 18, 2004) was a doctor of chemical sciences and a professor. He was in charge of the Division of Physical and Colloid Chemistry of the Tomsk Polytechnic Institute (from 1956) and was the scientific head of the problem laboratory at the same institute (from 1963). He graduated from the Ural Polytechnic Institute. In 1930, he began to engage in research work at the Institute of Chemistry and Metallurgy of the Urals Branch of the USSR Academy of Sciences (Sverdlovsk), defended his candidate thesis, and in 1951—his doctoral thesis. From 1951 Stromberg was associate professor, then professor of the Department of Physical and Colloid Chemistry of the Ural University (Sverdlovsk). He co-authored a textbook on physical chemistry (of which there were five editions). He led the Tomsk Seminar on Analytical Chemistry and worked in many councils and commissions.*

Fig. 4.4 Arman Genrikhovich Stromberg (September 16, 1910–September 18, 2004) performed a lot of work on the theory and applications of inversion voltammetry. Photo provided by A. G. Stromberg

Fig. 4.5 Kh'ena Zalmanovna Brainina (born May 26, 1930) authored work on the inverse voltammetry of solid phases, new electrodes, and determination of antioxidant activity. Picture provided by Kh. Z. Brainina

No less attention was paid to research in the field of SV of solid phases and surfaces (Kh. Z. Brainina (Fig. 4.5), E. Ya. Neiman, V. V. Slepushkin, A. I. Kamenev, N. Yu. Stozhko, et al.). Kh. Z. Brainina developed the SV of solid phases in several directions: (1) discharge–ionization of metals on an indifferent solid electrode—SV of metals; (2) electrochemical oxidation–reduction of ions of variable valence, accompanied by the formation of a poorly soluble compound on the surface of an indifferent electrode, and dissolution of these compounds—SV of ions of variable valence; (3) the formation, as a result of anodic polarization of the electrode, of poorly soluble compounds on the surface of the electroactive electrode and their reduction—SV of anions [21–23].

The practical applications of stripping methods have turned out to more diverse than the case of classical polarography. Among them, in particular, are applications for material analysis, determination of small impurity concentrations, including environmental samples, and the determination of antioxidant activity. The latter theme first appeared in the framework of coulometry with electrogenerated titrants–oxidizers; this generated interest among the specialists who developed the methodology of voltammetry on solid electrodes.

The results of experiment optimization and software and methodological support, as well as the information obtained during the mathematical modeling of analytical signals, helped to realize the concept of multi-element electrochemical analysis [18, 19, 24].

New electrodes, including those modified by various methods, were proposed, studied, and used for solving practical problems. O. A. Songina introduced the so-called mineral-paste electrodes [25]. O. L. Kabanova, in the 1970s, was among the first to suggest using a carbositall electrode for analysis [26, 27]. Much has been done in the field of modified electrodes, including work by G. K. Budnikov et al. [28]. One can note that the work of a number of electroanalysts on chemically modified electrodes was a continuation of the work undertaken in Russia at the turn of the century (Fig. 4.6).

German Konstantinovich Budnikov *was born on October 8, 1936. He graduated from the Kazan University. He is a doctor of chemical sciences and a professor. For many years he was in charge of the Department of Analytical Chemistry at the Kazan University—now the Kazan (Privolzhskii) Federal University. He is an honorary professor and member of the editorial boards of several journals. His interests are in the fields of organic analysis, electrochemistry, the chemistry of coordination compounds, ecomonitoring, biochemistry, and medicine. His successes were achieved when creating combined methods of the determination based on a combination of liquid extraction and voltammetry (polarography); electrochemical methods for determining organic substances in solutions using reagents arising on the electrode; methods for determining platinum metals with low detection limits on the basis of catalytic currents of hydrogen evolution in metal chelate solutions; sensors based on chemically modified electrodes with catalytic responses for stationary and flowing conditions for the determination of organic substances, including biologically active ones; amperometric biosensors based on immobilized enzymes and mercury film, as well as planar carbon electrodes, for the determination of environmental pollutants. He is also interested in the history of analytical chemistry in Kazan and electroanalytical chemistry in the USSR and*

Fig. 4.6 German Konstantinovich Budnikov developed voltammetric methods for determining organic compounds, including bioactive ones; he carried out research in the field of modified electrodes and biosensors. Picture provided by G. K. Budnikov

Russia. Professor Budnikov authored more than 600 articles and more than 20 books, and is one of the most cited Russian analysts (Fig. 4.7).

A logical continuation of research in the field of chemically modified electrodes have led to electrochemical biosensors—devices, in which a potentiometric or voltammetric transducer interfaces with a biochemical recognition element—an enzyme, antibodies, or oligonucleotides. The first enzyme electrode was created by L. Clark in 1962, and in Russia similar studies were carried out from the mid 1980s, most actively at the Division of Chemical Enzymology of the Moscow State University, but also at other scientific centers. In St. Petersburg, studies on potentiometric biosensors were carried out by E. B. Nikol'skaya. At the Kazan University, studies dealt with the development of methods for determining pesticides—cholinesterase inhibitors using amperometric devices. The first publication in this field was in the *Journal of Analytical Chemistry* in 1987. Research in the field of enzyme sensors was carried out at the A. N. Frumkin Institute of Electrochemistry of the Russian Academy of Sciences (RAS). Somewhat later, the

Fig. 4.7 Monograph on modified electrodes

topic of eletrochemical biosensors was taken on by the A. N. Bach Institute of Biochemistry of the RAS, and in the Division of Analytical Chemistry of the Moscow University by A.A. Karyakin.

4.4 Other Achievements in the Area of Electrochemical Methods

Quite a lot of effort was given to the development of coulometric methods; a review of Russian research in this field was published, as already mentioned, by Budnikov and Shirokova [3]. Among the original work was the creation of so-called sub-stoichiometric coulometry [29]. A large series of studies were devoted to the coulometric determination of noble metals. This work was carried out in the V. I. Vernadskii Institute of Geochemistry and Analytical Chemistry of the USSR Academy of Sciences and the N. S. Kurnakov Institute of General and Inorganic Chemistry of the RAS by O. L. Kabanova, N. A. Ezerskaya et al., as well as at the Kazan University. At the Vernadskii Institute, precision coulometers (using the potentiostatic version of the method), which provide a very high accuracy of the determination of macroelements, were created (V. A. Slavnyi, A. N. Mogilevskii, and I. G. Sentyurin). These instruments are manufactured and used in nuclear industry; at the Vernadski Institute, methods to precisely determine more than 15 elements were developed. At least two books have been devoted to coulometric analysis: Zozulya [30] and Agasyan and Khamrakulov [31], as well as a number of reviews and journal publications. Currently, at the Kazan University, coulometry with electrogenerated titrants is systematically used to determine individual antioxidants and to evaluate integrated indicators of complex objects of organic nature (G. K. Budnikov and G. K. Ziyatdinova) (Fig. 4.8).

Fig. 4.8 Precision coulometer, PIK-200, developed at the V. I. Vernadskii Institute of Geochemistry and Analytical Chemistry of the USSR Academy of Sciences. Photo taken from the website of the Vernadskii Institute of Geochemistry and Analytical Chemistry

The results of a thorough investigation of amalgams by M. T. Kozlovskii [32] were of importance for electrochemical production and, to a large extent, for chemical analysis. Several groups of researchers developed high-frequency methods of analysis, in particular high-frequency titration [33] and contact conductometry [34]. Amperometric titration methods have been widely applied [35]. Finally, we should remember that F. F. Reiss, a professor of the Moscow University, discovered electro-osmosis in 1808.

References

1. Scholz, F. (ed): Electrochemistry in a Divided World. Innovations in Eastern Europe in the 20th Century, 471pp, Springer, Heidelberg and oths. (2015)
2. Budnikov, G.K., Shirokova, V.I.: Zh. Anal. Khim. **64**, 1309 (2009)
3. Budnikov, G.K., Shirokova, V.I.: Zh. Anal. Khim. **70**, 92 (2015)
4. Compton, R.G.W., Gregory, G.Z., Zakharova, E.A., Stromberg, A.G.: First Class Scientists, Second Class Citizen: Letters from GULAG and History of Electroanalysis in the USSR, p. 380. Imperial College Press, London (2011)
5. Lubert, K.H., Kalcher, K.: Electroanalysis **22**, 1937 (2010)
6. Scholz, F.J.: Solid State Electrochemistry (2013). https://doi.org/10.1007/s10008-013-2107-2
7. Kryukova, T.A., Sinyakova, S.I., Aref'ieva, T.V.: Polarographic Analysis, 772pp. Goskhimizdat, Moscow (1959) (in Russian)
8. Lyalikov, Yu.S. (Ed.): Theory and Practice of Polarographic Analysis, 426pp. Shtiintsa, Kishinev (1962) (in Russian)
9. Nikolskii, B.P.: Acta Physicochim. **7**, 597 (1937)
10. Nikolskii, B.P., Tolmacheva, T.A.: Zh. Phys. Khim. **10**, 504 (1937)
11. Hayashi, K., Yamanaka, M., Toko, K., Yamafuji, K.: Sensors and Actuators **2**, 205 (1990)
12. Toko, K.: Mat. Sci. Eng. **4**, 69 (1996)
13. Vlasov, Y.G., Ermolenko, Yu.E, Legin, A.V., Rudnitskaya, A.M., Kolodnikov, V.V.: Zh. Anal. Khim. **65**, 900 (2010)
14. Vlasov, Y.G., Legin, A.V., Rudnitskaya, A.M.: Zh. Anal. Khim. **52**, 837 (1997)
15. Kalach, A.V., Zyablov, A.N., Selemenev, V.F.: Sensors in the Analysis of Gases and Liquids, 240pp, Voronezh (2011) (in Russian)
16. Kuchmenko, T.A.: Analitika i Kontrol **18**, 373 (2014)
17. Sidel'nikov, A.V., Maistrenko, V.N., Evtyugin, G.A.: Voltamperometric electronic tongue. In: Vlasov, Y.G. (ed) a book: Chemical Sensors (Problems of Analytical Chemistry, **14**), p. 285. Nauka, Moscow (2011) (in Russian)
18. Stradyn, Y.P., Mairanovskii, S.G. (eds): Polarography. Problems and Prospects, 416pp. Zinatne, Riga (1977) (in Russian)
19. Agasyan, P.K., Zhdanov, S.I. (eds.): Voltammetry of Organic and Inorganic Compounds, 248pp. Nauka, Moscow (1985) (in Russian)
20. Stromberg, A.G., Kaplin, A.A., Karbainov, Y.A., Nazarov, B.F., Kolpakova, N.A. Slepchenko, G.B., Ivanov, Y.A., Izv. Vissh. Uchebn. Zaved., Khimia i Khim. Technol. **43**, 10 (2000)
21. Brainina, Kh: Stripping Voltammetry in Chemical Analysis, p. 222. Wiley, New York (1972)
22. Brainina, Kh, Neyman, E.: Electroanalytical Stripping Methods, p. 198. Wiley, New York (1993)
23. Brainina, K.Z., Neiman, E.Y., Slepushkin, V.V.: Inversion Electroanalytical Methods, 240pp. Khimiya, Moscow (1988) (in Russian)
24. Kamenev, A.I., Viter, I.P.: Ross. Khim. Zh. (Zh. Ross. Khim. Ob.), **1**, 77 (1996)

25. Songina, O.A.: Talanta **25**, 116 (1978)
26. Kabanova, O.L., Goncharov, Y.A.: Zh. Anal. Khim. **28**, 1665 (1973)
27. Anokhin, B.A., Ignatov, V.I.: Zh. Anal. Khim. **29**, 1221 (1974)
28. Budnikov, G.K., Evtyugin, G.A., Maistrenko, V.N.: Modified Electrodes for Voltammetry in Chemistry, Biology, and Medicine, 416pp. BINOM. Lab. Znanij., Moscow (2009) (in Russian)
29. Agasyan, P.K., Khamrakulov, T.K.: Zh. Anal. Khim. **23**, 19 (1968)
30. Zozulya, A.P.: Coulometric Analysis, 160pp. Khimiya, Leningrad (1968) (in Russian)
31. Agasyan, P.K., Khamrakulov, T.K.: Coulometric Method of Analysis, 168pp. Khimiya, Moscow (1984) (in Russian)
32. Kozlovskii, M.T.: Selected Works. I. Analytical Chemistry. Electrochemical Methods of Analysis. II. Electrolysis with Mercury Cathode and Amalgams Carburizing, Alma-Ata, 1974, I, 303pp, II, 324pp. (in Russian)
33. Zarinskii, V.A., Ermakov, V.I.: High-frequency Chemical Analysis, 200pp. Nauka, Moscow (1970) (in Russian)
34. Griliches, M.S., Filanovskii, B.K.: In: Aguf, I.A. (ed.) Contact Conductometry, 176pp. Khimiya, Leningrad (1980) (in Russian)
35. Songina, O.A., Zakharov, V.A., Amperometric Titration, 3rd ed., 304pp. revised. Nauka, Moscow (1979) (in Russian)

Chapter 5
Chemical Methods of Analysis

Abstract Several widely used analytical reactions were proposed in Russia, such as the reaction of phosphate with ammonium molybdate (G. V. Struve) or the reaction of nickel with dimethylglyoxime (L. A. Chugaev). N. A. Tananaev and F. Feigl simultalneously developed drop analysis. A great contribution was made to the theory of the action of complexing organic reagents on metal ions, and new such reagents were created (V. I. Kuznetsov, S. B. Savvin, etc.); particularly widespread was the reagent Arsenazo III. The kinetic method of analysis owes its generation to K. B. Yatsimirskii.

5.1 General Remarks

For several centuries, chemical methods have been the dominant methods used for chemical analyses. Excluding the famous analysis of the gold crown of King Hieron, by Archimedes, physical methods began after the development, in 1859, of spectral analysis by Bunsen and Kirchhoff. At approximately the same time, the use of electrolysis was developing, a method which may be considered physicochemical. Until the middle of the twentieth century, chemical methods played a huge role, sometimes experiencing a steep rise in use, as was the case during the widespread distribution of complexometric titration in the postwar years, and the creation of automatic titrimeters, used to determine nitrogen, according to Kjeldahl, or water, according to Fischer.

It is well known that chemical methods are now very widely used practically, but they have gradually ceased to be the subject of large-scale scientific research. These methods are studied in the courses of analytical chemistry at universities, because, among other things, they make it possible to deepen students' understanding of chemistry as a result of making them acquainted with chemical reactions, their kinetics and thermodynamics, chemical equilibrium, stability and other characteristics of coordination compounds, etc. In addition, it is important that every chemist should be able to determine the concentration of acid by titration, for example.

Y. A. Zolotov, *Russian Contributions to Analytical Chemistry*,
https://doi.org/10.1007/978-3-319-98791-0_5

In Russia, of course, many researchers were engaged in the development and improvement of chemical methods of analyses. Several reactions proposed in the nineteenth and early twentieth centuries were among those widely used, such as the reaction of phosphate with ammonium molybdate, cobalt with 1-nitroso-2-naphthol, or nickel with dimethylglyoxime. Methods of obtaining pure precipitates for gravimetry were studied, e.g., by N. A. Tananaev. He, along with the Austrian chemist Fritz Feigl, developed drop analysis. It is necessary to note the introduction into practice of effective organic reagents for the photometric determination of elements, as discussed in Chap. 1.

A large volume of work dealt with titrimetric methods, in particular, to non-aqueous titration. Non-standard titrants were suggested by A. I. Busev (divalent chromium compounds) and V. S. Syrokomskii (vanadatometry). Interesting modern approaches to titrimetry were developed by B. M. Mar'yanov.

It is especially necessary to note the formation of the kinetic method by K. B. Yatsimirskii. Although some kinetic methods were developed before his, Yatsimirskii created the system, proposed original techniques for implementing the methods, and developed many specific procedures.

5.2 Some Chemical Reactions

The reaction of phosphate with ammonium molybdate, widely used to determine phosphorus, leading to the formation of a heteropoly compound, was first described in 1848 by Struve and Svanberg [1]. Later, Struve found that arsenic gave a similar reaction; a method for detecting this element was also developed (Fig. 5.1).

***Genrikh Vasil'evich Struve** (1822–1908) was a corresponding member of the St. Petersburg Academy of Sciences, worked as an assayer in the Department of Mining and Salt Mines in St. Petersburg, and was an expert in forensic chemistry in the Caucasus. He developed a method for the determination of very small amounts of hydrogen peroxide with lead dioxide (1869). He published tables for calculating*

Fig. 5.1 Genrikh Vasil'evich Struve (1822–March 22, 1908) discovered the reaction of phosphate with ammonium molybdate. Photo previously published

Table 5.1 Some analytical reactions and reagents proposed by Russian chemists

Analit	Reaction, reagent	Author	Year	References and comments
Phosphorus	Ammonium molybdate	G. V. Struve, G. Svanberg	1848	[1]
Aromatic hydrocarbons	β-Dinitroanthraquinone	Yu. F. Fritzsche	1863	[2] ("Fritzsche Reagent")
Halogens in organic compounds	Green flame when heated in the presence of copper oxide	F. F. Beilstein	1872	[2] ("Beilstein test")
Copper, nickel, and cobalt	Xanthogenates	P. N. Akhmatov	1870	[3]
Cobalt	1-Nitroso-2-naphthol	M. A. Il'inskii, G. Knorre	1884	[4]
Mobile hydrogen in organic compounds	–	L. A. Chugaev, F. V. Tserevitinov	The end of the nineteenth century	[5] ("Tserevitinov method")
Nickel	Dimethylglyoxime	L. A. Chugaev	1905	[6]
Uranium	Thoron (Thorin, Arsenazo)	V. I. Kuznetsov	1941	[7]
Many metal ions	Arsenazo III	S. B. Savvin	1959	[8]

the results of quantitative analysis; for a long time, the tables were used by laboratory workers (G. V. Struve, Chemical Tables Serving for the Calculation of Quantitative Decompositions, St. Petersburg, 1853, in Russian).

Several other reactions proposed for the determination of elements and certain compounds are presented in Table 5.1.

The determination of elements using organic reagents will be discussed in more detail below.

N. A. Tananaev formulated, in the 1920s, rules for obtaining clean precipitates for gravimetric analysis. At the same time, and in parallel with F. Feigl (Austria), he developed drop analysis [9]; later, he created the so-called chip-free analysis of metals and alloys method [10] (Fig. 5.2).

Nikolai Aleksandrovich Tananaev *[May 06(18), 1878–June 7, 1959] was a doctor of chemical sciences, a professor, an Honored Scientist, and laureate of the State Prize of the USSR (1949). He was a head of the Division of Analytical Chemistry at the Kiev Polytechnical Institute (1921–1938) and the Urals Industrial (Polytechnical) Institute (1938–1959). From 1919, he began to develop the drop method, which was a method of qualitative analysis. Detection of ions in a complex mixture was carried out on a drop of the test solution on filter paper in the form of colored spots or rings. The interfering ions were not previously separated, and*

Fig. 5.2 Nikolai Aleksandrovich Tananaev developed a drop analysis method, a method of brushless analysis. Photo previously published

their influence was eliminated from the same drop on the paper by the action of reagents. Later, he successfully implemented the idea of detecting ions in a complex mixture by developing a fractional analysis method performed in test tubes. The results of the studies were summarized in the book Drop Analysis, the first edition of which was published in 1926. A significant part of N. A. Tananaev's research was devoted to the improvement of gravimetric analysis. The methods of precipitation of crystalline and amorphous sediments, proposed by him, have found application in analytical practice. The results of research in this field are set forth in the book The Weight Analysis, the first edition of which was published in 1931. In 1938, N. A. Tananaev moved to Sverdlovsk. During World War II, he began developing a rapid method for analyzing trophy weapons. Initially, it was necessary to establish the chemical composition of alloys, from which the weapon was made, without destroying their integrity. This was the origin of the chip-free method for the analysis of alloys of ferrous, non-ferrous, and noble metals (later this method was also used to analyze glass products). For this work he was awarded the State Prize of the USSR.

5.3 Determination of Elements Using Organic Reagents

Organic analytical reagents are organic compounds or their combinations which chemically interact with analytes and used (due to this interaction) for detection or determination of these analytes. The analytes may be cations or anions, inorganic or organic compounds, or separate functional groups. Reagents are used as indicators, titrants, extractants, they are used in electrochemical and chromatographic methods, but more often in photometric analysis. The works of Stephen [11], Fedorenko [12], Strel'nikova [13, 14], etc., are devoted to the history of the OAR.

Up to the end of the eighteenth century, natural organic substances were used as organic reagents. Probably the first documented use of an organic substance for the recognition of "analytes" is Pliny's description of the use of papyrus impregnated with extract of tanning nuts (outgrowths on the undersides of oak leaves) in order to distinguish between iron and copper salts. Iron gives a color reaction with the derivatives of tannin contained in tanning nuts, forming a black compound; copper causes green coloration. The following written evidence of the use of natural organic substances for chemical analysis purposes dates back to the sixteenth–seventeenth centuries. The medical doctor and pharmacist O. Tahheny described, in 1666, the reactions of aqueous extracts of tanning nuts with solutions of the salts of several metals. Stephen [11] believes that this was the first systematic study of reagent reactions with metal ions. Other OARs dating back to the sixteenth–seventeenth centuries, in addition to extracts of tanning nuts, were plant extracts, some of which were used for dyeing fabrics. R. Boyle also actively studied extracts of violets, cornflower, and litmus. One of the substances used by Boyle was the first fluorescent indicator for acid (fluorescence disappeared with the addition of acid). The juice of violets subsequently served as the basis for the indicator used for acid–base titration, which was first used in the second half of the eighteenth century. Litmus and turmeric were also used for this purpose (K. F. Wenzel, 1777).

In 1776, K. Scheele synthesized oxalic acid and demonstrated its reaction with calcium. Apparently, oxalic acid was the first synthetic OAR. In the first half of the nineteenth century a number of organic compounds were used as indicators for titrimetry. In 1870, P. N. Akhmatov proposed xanthogenates as reagents for cadmium and other elements [3]; in 1884, M. A. Il'inskii discovered the reaction of cobalt with 1-nitroso-2-naphthol [4]; and in 1905, L. A. Chugaev described dimethylglyoxime as a reagent for nickel [6]. Griss created a reagent for the determination of nitrite. However, the most important was the use of organic compounds as colorimetric reagents for metal ions and for use as titrants (complexonometry, G. Schwarzenbach, 1940–1950).

Photometric (initially colorimetric) methods of determining elements became very widespread in the first half of the twentieth century. The inorganic reagents used for this purpose (thiocyanate, ammonia, etc.) were gradually replaced by organic ones, which first appeared as a result of almost random observations, and then taking into account the generalizations formed, the regularities. Taking into account the diversity of the OARs in terms of their chemical nature and areas of use, it should be concluded that there cannot be a general theory of the action of OARs. However, in terms of complexing reagents, most often used to determine the elements, such a theory is not only possible, but is well developed; a significant contribution to its formation was made by N. S. Kurnakov, L. A. Chugaev, V. I. Kuznetsov, L. M. Kul'berg, S. B. Savvin, and other Russian chemists–analysts. So, N. S. Kurnakov, L. A. Chugaev, and L. M. Kul'berg developed ideas about the characteristic groupings of atoms in molecules of complexing reagents. In the text below, based on the article by N. V. Fedorenko [12], the idea of characteristic atomic groupings is covered.

In his work *On Complex Metal Bases* (1893) [15] N. S. Kurnakov first suggested that the ability of certain groups of organic compounds to react with a certain metal depends on the presence, in their molecules, of a combination of atoms common to a given group of reagents. Later this suggestion became the main idea on which the theory of the action of organic complexing reagents developed—the fact that it was first expressed by Kurnakov was forgotten. In 1893, Kurnakov studed the interaction of thiourea with platinum salts and mad the conclusion that the interaction was caused by complexation, platinum was directly connected with thiourea molecules. He expressed the opinion that the reactivity of thiourea and its derivatives is determined by the presence of a certain grouping of atoms in their molecules. Kurnakov wrote: "The ability to combine thioureas depends on the presence of the CS group in them" [15, p. 73]. In the same paper, he suggested that other organic compounds containing this group should interact with platinum salts like thiourea and form similar reaction products. "If the ability of thiourea to combine with metals is due to the presence of $CSNH_2$ or C(NH)·SH groups," Kurnakov wrote, "then it is obvious that thioamides and thiourethanes should have the same properties" [15, p. 78]. This proposal was confirmed experimentally by Kurnakov. Platinum compounds with both thioamides and thiourethanes were obtained, with appearance and properties similar to those of thiourea.

However, for researchers involved in organic reagents, Kurnakov's conclusion went unnoticed. The reason, apparently, was that Kurnakov's research was devoted to the study of complex compounds, and the action of OARs at the time was not associated with the formation of complexes between organic reagents and metals.

L. A. Chugaev (1873–1922), a prominent scientist who created the school of researchers of complex compounds, was, apparently, the first to draw attention to this connection. In 1905, as already mentioned, he proposed dimethylglyoxime for the determination of nickel [6]. Chugaev found that the ability of oximes to interact with nickel is due to the presence in their molecules of a certain combination of atoms of two =NOH groups at neighboring carbon atoms. Characterizing oximes as analytical reagents, Chugaev wrote: "These compounds are characterized by the presence of a kind of atomic grouping, in which the oximid groups partly exchange their hydrogen for the metal and function as acid residues, partly play a role analogous to the role of ammonia in metal-ammonium compounds" [16, p. 490] (Fig. 5.3).

Thus, both N. S. Kurnakov and L. A. Chugaev showed that the ability of an organic compound to interact with an element is due to the presence in its molecule of a specific form of a given metal group of atoms, through which interaction takes place. However, N. S. Kurnakov made this conclusion about 15 years before L. A. Chugaev. The groups that Kurnakov and Chugaev talked about in modern literature were called "characteristic atomic groups." This name was introduced in 1925 by Feigl [17, 18]. However, sometimes the very discovery of these groups was attributed to him. The theory of characteristic atomic groups was developed and systematized by Kul'berg [19].

V. I. Kuznetsov proposed several general approaches for the direct acquisition of organic complexing agents, e.g., the so-called "analogy hypothesis" (analogies of the action of organic and inorganic reagents—ligands having identical atoms directly interacting with metal ions) [20]. One hypothesis considered a qualitative parallel between the conditions for carrying out reactions with organic reagents and simple inorganic reactions (hydrolysis, precipitation of sulfides, and formation of

Fig. 5.3 Lev Aleksandrovich Chugaev (October 16, 1873–September 22, 1922) contributed to the study of metal chelates and proposed dimethylglyoxime as a reagent for nickel. Photo previously published

ammoniates) containing the corresponding donor atoms in the functional analytical groups. This hypothesis made it possible to predict ways of analytical using reagents.

In the USSR extensive research was launched aimed at creating new organic reagents; at the same time, the previously proposed reagents—dithisone, dithiocarbamates, 8-hydroxyquinoline, etc.—were also used on a large scale. The creation and investigation of complexing reagents relied on the theoretical basis discussed above. Among the producers of new organic reagents for sensitive and/or selective determination of metal ions should be mentioned V. I. Kuznetsov, S. B. Savvin, A. M. Lukin, S. I. Gusev, Yu. M. Dedkov, and N. N. Basargin.

V. I. Kuznetsov introduced several reagents, which he produced himself, into the practice of analysis. Thus, the reagent thoron was distributed to many countries; it was used to determine uranium and thorium [7]. The reagent was included in the catalogs of a number of companies that produced chemical reagents. Among the other reagents produced by V. I. Kuznetsov was arsenazo, or uranone (later it became known as arsenazo I), which was similar in its structure, its properties, and its applications to thoron, stilbazo, and other reagents (Fig. 5.4).

Vitalii Ivanovich Kuznetsov *(May 4, 1909–October 1, 1987) worked at the All-Union Institute of Mineral Raw Materials (VIMS), headed at one time the Division of Analytical Chemistry at the M. V. Lomonosov Moscow Institute of Fine Chemical Technology, and then for a long time headed the laboratory of organic reagents at the V. I. Vernadskii Institute of Geochemistry and Analytical Chemistry of the USSR Academy of Sciences. Subsequently, he was an employee of the Central Institute of Agrochemical Agricultural Services (CINAO).*

An employee of V. I. Kuznetsov, the former his student professor Sergei Borisovich Savvin synthesized arsenazo III [8], a reagent that has become one of the most common. It was produced in several countries and used for the

Fig. 5.4 Vitalii Ivanovich Kuznetsov developed the theory of the action of organic analytical reagents on metal ions and proposed a series of effective reagents.
Picture provided by V.V. Kuznetsov

photometric determination of rare-earth elements, zirconium, uranium, plutonium, and a number of other elements. A monograph was devoted to the reagent and its analogs [21] (Figs. 5.5 and 5.6).

Here is a incomplete list of new reagents which have become widespread:

Diantypyrylmethane—S. I. Gusev, 1950.
Thoron—V. I. Kuznetsov, 1941.
Beryllon II—A. M. Lukin and G. V. Zavorikhina, 1950.
Stilbazo—V. I. Kuznetsov, G. G. Karnovich, and D. A. Drapkina, 1950.
Arsenazo I—V. I. Kuznetsov, 1955.
Arsenazo III—S. B. Savvin, 1959.
Lumagallion—A. M. Lukin and E. A. Bozhevol'nov, 1960.
Carboxyarsenazo—K. F. Novikova, N. N. Basargin, and Tsuganova, 1961.
Pikramin E—Yu. M. Dedkov, 1970.
Sulfochlorphenol S—S. B. Savvin, 1964.

Organic photometric reagents created by N. N. Basargin (Nitchromazo, Carboxyarsenazo, Dichlorchromotropic acid, Arsenazohimdu, Stilbazohimdu, etc.) have been used, and some reagents have entered the catalogs of firms which produce reagents (Fig. 5.6).

Organic reagents created by V. I. Kuznetsov et al. contributed to the solution of major national economic and defense tasks in 1940s–1960s. Particularly well known are reagents belonging to the arsenazo and thoron groups. Both the arsenazo reagent and the reagents with the doubled main grouping (Arsenazo III, etc.), made it possible to solve some analytical problems in the nuclear industry; in particular, highly

Fig. 5.5 Reagent Arsenazo III

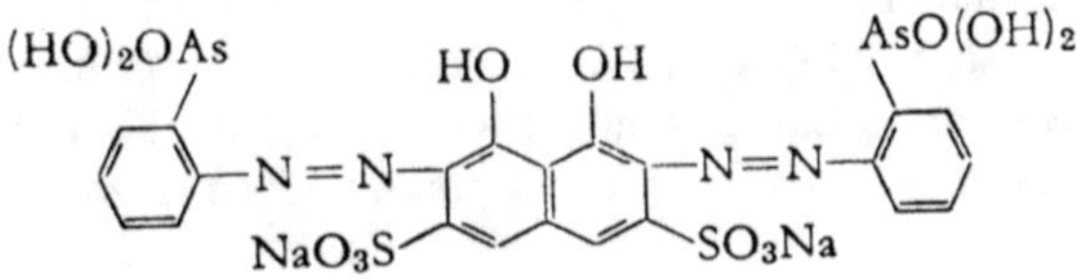

$(HO)_2OAs$ HO OH $AsO(OH)_2$
$-N=N-$ $-N=N-$
NaO_3S SO_3Na

Fig. 5.6 Sergei Borisovich Savvin (January 1, 1931–July 6, 2014) was a specialist in the field of organic analytical reagents, especially photometric analysis and preconcentration, and created the well-known reagent Arsenazo III. Picture provided by the Vernadskii Institute of Geochemistry and Analytical Chemistry

sensitive photometric methods were developed for the determination of uranium, thorium, and other elements accompanying uranium raw materials and nuclear fuel. Theoretical and practical issues linked to obtaining and identifying transplutonium elements, control issues in metallurgy, processing mineral raw materials, medicine, agriculture, controlling environmental pollution, and other areas, were solved.

5.4 Development of Titrimetric Methods

In the middle of the last century, new titrimetric methods were proposed and developed, mainly redox methods. A. I. Busev developed chromometry, a method for the determination of oxidants, based on the use of chromium (II) compounds as a titrant-reductant. The analysis is carried out in an acidic medium; the end point is determined potentiometrically, amperometrically with a rotating platinum microanode, or using chemical redox indicators. Many procedures of titration have been developed, e.g., for the determination of cerium (IV), arsenic (V), and a number of organic compounds [22].

V. S. Syrokomskii created vanadatometry, a method for the determination of reducing agents using vanadium (V) compounds as a titrant–oxidant; vanadium in this case is reduced to vanadium (IV) [23]. A lot of concrete procedures for such a determination have been created. Indicators include diphenylamine and *p*-phenylarsonic acid. A greater selectivity is achieved compared with, e.g., chromatometry or cerimetry. It is possible, e.g., to determine hydroquinone in the presence of cresols or maleic acid in the presence of formic acid.

The research of B. M. Mar'yanov is interesting [24, 25]. Using modern techniques, especially chemometric ones, he was able to substantially increase the total potential of titrimetry, e.g., Mar'yanov showed the possibility of significantly advancing the range of concentrations determined by titrimetry methods toward low values (Fig. 5.7).

Fig. 5.7 Boris Mikhailovich Mar'yanov improved the titrimetric method of analysis. Photo provided by B. M. Mar'yanov

Boris Mikhailovich Mar'yanov *(April 4, 1932–February 6, 2006) was a doctor of chemical sciences and a professor. He was in charge of the Division of Analytical Chemistry at the Tomsk University. He graduated (1955) from the Chemistry Department of the Tomsk University where he worked as assistant professor at the Division of Analytical Chemistry. Mar'yanov worked on the theory of titration and chemometrics. He authored 140 works, including 3 books. Under his leadership, 9 candidates of sciences were trained.*

5.5 Kinetic Methods

Separate procedures of analysis, which are kinetic by their mechanism, were described in the nineteenth century and in the first half of the twentieth century, e.g., one such procedure was developed by well-known American analysts Kolthoff and Sandell [26]. The history of such determinations is covered briefly in a book written by Zolotov and Vershinin [27]. However, as a universal method of analysis, the kinetic method was actually developed by K. B. Yatsimirskii.

K. B. Yatsimirskii, while working at the Ivanovo Chemistry and Technology Institute formulated the main theoretical basis of the method, substantiated criteria for its application, demonstrated the technique's possibilities and limitations, and developed experimental approaches. He showed that the use of the rate of the catalytic chemical reaction, as an analytical signal, makes it possible to determine amounts of catalysts, with very low detection limits, without the need for expensive and complex equipment. Different variants of the kinetic method, primarily catalytic ones, were proposed, namely, the determination directly from the reaction rate using the integral or differential form of the kinetic equation, determination by duration of the induction reaction period, catalymetric titration, and polarographic catalytic currents. One profound study of kinetics, and the mechanism of the reactions used in kinetic methods, became the basis for improving methods by introducing activators.

Under the guidance, and with the direct participation, of K. B. Yatsimirskii his trainees developed kinetic methods for determining a large number of elements. Many of the proposed procedures differ not only in their low detection limits, but also in their high selectivity, which makes it possible to use them for the analysis of natural materials or the creation of technological solutions. In K. B. Yatsimirskii's monograph *Kinetic Methods of Analysis* the application of kinetic regularities in the analysis was justified for the first time. This book was published twice in Russian [28, 29], as well as being translated into English, Polish, and Hungarian (Fig. 5.8).

***Konstantin Borisovich Yatsimirskii** (April 4, 1916–June 21, 2005) was an academician of the National Academy of Sciences of Ukraine, a doctor of chemical sciences, and a professor. He was in charge of the Division of Analytical Chemistry of the Ivanovo Chemical Technology Institute and, later, director of the L. V. Pisarzhevskii Institute of Physical Chemistry of the Ukrainian SSR Academy of Sciences. He graduated (1941) from the Central Asian University in Tashkent, and immediately after graduation defended, in the same year, his candidate of sciences dissertation. During World War II, he taught military chemistry at the Podolsk Infantry School. In 1945–1961, he worked at the Ivanovo Chemical Technology Institute, where he worked his way from an assistant to the head of the Division of Analytical Chemistry and Deputy Director of the Institute. In 1948, he defended his doctoral dissertation, in 1961 he was elected a corresponding member, and in 1964—a full member of the Ukraine Academy of Sciences. From 1962 he worked in the Academy of Sciences of the Ukrainian SSR in Kiev—he headed departments in the Institute of General and Inorganic Chemistry, and then the L. V. Pisarzhevskii Institute of Physical Chemistry, for which he was director from 1969–1982* (Fig. 5.9).

K. B. Yatsimirskii was an outstanding chemist in the field of inorganic and analytical chemistry. His work on inorganic and coordination chemistry (thermochemistry and thermodynamics of coordination compounds, their electronic structure and structural parameters, kinetics and catalysis of reactions involving

Fig. 5.8 Konstantin Borisovich Yatsimirskii (April 4, 1916–June 21, 2005) made a decisive contribution to kinetic methods of analysis. Photo previously published

Fig. 5.9 The first edition of the 1963 monograph by K. B. Yatsimirskii *Kinetic Method of Analysis*

coordination compounds, and coordination compounds with macrocyclic ligands and bioligands) is of the greatest importance. Practically all the results of research in these areas are to some extent related to analytical chemistry—the use of new information obtained in his work on the properties of coordination compounds in various methods of determination. The main achievements of K. B. Yatsimirskii in analytical chemistry are related to kinetic methods of analysis, the development of which he founded.

K. B. Yatsimirskii was known as a brilliant lecturer, a popularizer of scientific knowledge. He was an Honored Worker of Science and Technology and laureate of the State Prize of Ukraine and prizes named after L. A. Chugaev of the USSR Academy of Sciences and L. V. Pisarzhevskii of the National Academy of Sciences of Ukraine. He was awarded the Ya. Geirovskii Gold Medal of the Czechoslovak Academy of Sciences, elected *doctor honoris causa* of Wroclaw University, an honorary member of the Polish Chemical Society, and a corresponding member of

Academia Peloritana dei Pericolanti (Italy). He authored more than 1000 publications, including 24 monographs and 4 textbooks. Among his students were 17 doctors and 59 candidates of sciences.

5.6 Other Chemical Methods

Complexing analytical reagents, containing on the periphery of a molecule a group that is a stable free radical (see Chap. 1), were synthesized and studied. The idea was that the path of the reagent labeled in such a way can be tracked using the electron magnetic resonance method. Reagents were used for extraction-radio spectroscopic (EPR) determination of metal ions bound into complexes due to the complexing group of the reagent. Difficulties arose at the stage of separation of the complex from the excess of the reagent, which, after all, also gives an EPR signal. Nevertheless, a lot of work was done in this direction, and a large review was published [30].

References

1. Struve, G.V., Svanberg, L.Z.: Pract. Chemie **44**, 291 (1848)
2. Zolotov, Y.A.: Who was Who in Analytical Chemistry in Russia and the USSR, 3rd ed., 400pp. URSS, Moscow (2017) (in Russian)
3. Akhmatov, P.N.: On the Reactions of Xanthogenates with Compounds of Some Metals, Moscow (1874) (in Russian)
4. Il'inskii, M.A., Knorre, G.: Ber. Dtsch. Chem. Ges. **17**, 2581 (1884)
5. Volkov, V.A., Kulikova, M.B.: Russian Professorship. XVIII–early XX Century. Chemical sciences. Biographical dictionary, 275pp. Publ. House of the Russian State University of Architecture and Civil Engineering, St. Petersburg (2004) (in Russian)
6. Tschugaeff, L.: Ber. Dtsch. Chem. Geselsch. **38**, 2520 (1905)
7. Kuznetsov, V.I.: Dokl. AN SSSR **31**, 895 (1941)
8. Savvin, S.B.: Dokl. AN SSSR **127**, 1231 (1959)
9. Tananaev, N.A.: The Drop Method, 275pp. Goskhimizdat, Moscow and Leningrad (1954) (in Russian)
10. Tananaev, N.A.: A Chip-Free Method for Analyzing Black, Colored, and Precious Alloys, 210pp. Metallurgizdat, Sverdlovsk and Moscow (1948) (in Russian)
11. Stephen, W.I.: The Analyst **102**, 793 (1977)
12. Fedorenko, N.V.: Vopr. Istorii Estestv. Tekhniki **1**, 89 (1982)
13. Strel'nikova, E.B.: Vopr. Istorii Estestv. Tekhniki. **4**, 48 (1986)
14. Savvin, S.B., Strel'nikova, E.B.: Zh. Anal. Khim. **38**, 727 (1983)
15. Kurnakov, N.S.: Ref. by [12]
16. Chugaev, L.A.: Selected Works, in 3 vol. Izd. AN SSSR, Moscow (1954–1962) (in Russian)
17. Feigl, F.: Chemistry of Specific, Selective and Specific Reactrons, 550pp (1950)
18. Feigl, F.: Monatsh. **45**, 115 (1924)
19. Kul'berg, L.M.: Organic Reagents in Analytical Chemistry, Moscow and Leningrad: Goskhimizdat, 264pp. (1950) (in Russian)
20. Kuznetsov, V.I.: Zh. Anal. Khim. **2**, 67 (1947)

21. Savvin, S.B.: Arsenazo III. Methods for the Photometric Determination of Rare and Actinide Elements, 256pp. Atomizdat, Moscow (1966) (in Russian)
22. Busev, A.I.: Application of Divalent Chromium Compounds in Analytical Chemistry, 161pp. VINITI, Moscow (1960) (in Russian)
23. Syrokomskii, V.S., Klimenko, Y.V.: Vanadatometry. A New Method of Volumetric Chemical Analysis, 171pp. Metallurgizdat, Sverdlovsk and Moscow (1950) (in Russian)
24. Mar'yanov, B.M.: Method of Linearization in Instrumental Titrimetry, 158pp. Izd. Tomsk. un-ta, Tomsk (2001) (in Russian)
25. Mar'yanov, B.M., Chashchina, O.V.: Calculations of Ionic Equilibria. Handbook on Analytical Chemistry, 3rd ed., 152pp, revised and addit. Izd. Tomsk. un-ta, Tomsk (2006) (in Russian)
26. Sandell, E.B., Kolthoff, I.M.: J. Am. Chem. Soc. **56**, 1426 (1934)
27. Zolotov, Y.A., Vershinin, V.I.: History and Methodology of Analytical Chemistry, 464pp. ITs Academiya, Moscow (2007) (in Russian)
28. Yatsimirskii, K.B.: Kinetic Methods of Analysis, 190pp. Goskhimizdat, Moscow (1963) (in Russian)
29. Yatsimirskii, K.B.: Kinetic Methods of Analysis, 2nd ed., 204pp, correct. and addit. Khimiya, Moscow (1967) (in Russian)
30. Nagy, V.Yu., Petrukhin, O.M., Zolotov, Yu.A: CRC Crit. Rev. Anal. Chem. **17**, 265 (1987)

Chapter 6
Separation and Preconcentration Methods (Excluding Chromatography)

Abstract A general methodology for analytical preconcentration has been developed. Much has been done in liquid–liquid extraction, in particular the extraction of chelates and complex acids; the mutual influence of the elements during extraction has been discovered and studied (Yu. A. Zolotov). Effective sorbents for the preconcentration of ions of elements have been created (G. V. Myasoedova, G. I. Tsizin and others). Capillary isotachophoresis has been invented.

6.1 General Remarks

Except for the attempts of the alchemists to turn various metals into gold and create an "elixir of life," the prehistory of chemistry is to a considerable extent the history of the decomposition of natural substances and the isolation of individual components. Targeted synthesis was developed with a significant lag and emerged in priority positions only in the nineteenth century, with the rise of organic chemistry. In classical qualitative chemical analysis, especially systematic, the separation of substances by successive deposition occupied a key position. Quantitative analysis by chemical methods did not always require the obligatory separation of mixtures (titrimetry), but nevertheless, in this case the operations in question played, and play, an important role. With the development of instrumental methods of analysis, separation of mixtures and the preconcentration of small quantities began to acquire in some measure the nature of auxiliary, although very essential, and often necessary.

Separation and preconcentration became integral parts within the development of hybrid methods of analysis. We see this in modern chromatography (separation plus detection in one instrument), capillary electrophoresis, and in a number of variants of flow-injection analysis. And in such methods as mass spectrometry or ion mobility spectrometry, the separation of mixtures (more precisely, ions of the corresponding components of mixtures) is present "inside" the method. In summary, in terms of chemical analysis, the operations under consideration here are undoubtedly very important.

Y. A. Zolotov, *Russian Contributions to Analytical Chemistry*,
https://doi.org/10.1007/978-3-319-98791-0_6

Numerous methods serve the purposes of analytical separation and concentration; in scientific and methodological terms they are developed to different degrees and are very different in terms of the scale of their use. In addition, there are constantly emerging new methods to add to an analysts tool kit. Objects upon which these methods can be applied are extremely diverse; it is easier to say that these can be applied to almost any object.

The separation methods used in analytical chemistry are also used in radiochemistry; techniques used in this case—in general, are similar too. An obvious example is liquid–liquid extraction; there are many analysts using this method and many radiochemists. The similarity is confirmed by the fact that many chemists work simultaneously in areas of "analytical" and "radiochemical" separation; this particularly applies to the development of methods for the separation of elements.

To preconcentrate the microquantities, the same methods are usually used as for the separation of mixtures, but not all of them, and not all the time. So, chromatography is rarely used for preconcentration, and directional crystallization and zone melting—for the separation of mixtures. Interest in specific methods of analytical separation and preconcentration has changed over time. The peak of interest in methods shifted from precipitation (in part coprecipitation) to ion exchange, and especially to liquid–liquid extraction, then to sorption (solid phase extraction); in the case of the separation of mixtures—from precipitation and distillation to chromatography, with the latter being presently triumphant.

In the USSR, a general methodology of analytical preconcentration was formulated, aimed mainly at the preliminary concentration of microelements. In the 1950s–1980s, in line with the rest of the world, the USSR paid much attention to liquid–liquid extraction; several groups worked very actively in this direction, with a lot of theoretical work being undertaken along with the creation of applied techniques, new extractants, and a concerted effort in terms of analytical applications. Then the center of interest began to shift toward sorption methods. Of the other methods in use, we should be aware of capillary isotachophoresis.

6.2 Methodology of Analytical Preconcentration

The preconcentration of microcomponents before their determination occupies an important place in analytical chemistry. This is determined by the insufficient sensitivity of the methods of direct determination; the desire to deal with the same type of samples during determination, in order to facilitate graduation; and the inability to directly analyze samples, in which the desired components are distributed non-homogeneously, such as gold in rocks. Originally developed to be applied to microelements, analytical preconcentration is now very widely used in the analysis of organic substances and bio-objects, especially in combination with chromatographic determination (solid-phase microextraction, liquid microextraction, etc.).

A general methodology for analytical preconcentration has been developed [1–3]. It includes a number of statements, recommendations, assessments, etc., some of which seem obvious and known; others, perhaps, appear original, but somewhat controversial. However, on the whole, individual statements, generalizations, etc., form a system which can be considered as a methodology. Here are its items:

1. A definition of analytical preconcentration and its division into absolute and relative, group and individual.
2. The quantitative characteristics of preconcentration.
3. The advantages and disadvantages of preconcentration.
4. Appropriate (an inappropriate) uses of preconcentration.
5. Comparison of preconcentration with other methods.
6. An assessment of the rationality of combinations of preconcentration methods with methods of subsequent determination, along with isolation of the most successful combinations—the concept of hybrid and combined methods (Fig. 6.1).

In 1977, the concept and the term "hybrid methods of analysis" were proposed [4, 5]. These are methods of chemical analysis, based on the combination of separation and concentration of substances, on the one hand, and their determination (detection), on the other. Such a combination can be so dense that a new method actually appears, which is both a separation method and a method of determination. Examples are capillary electrophoresis and almost all variants of modern chromatography, in which the separation of a mixture of substances in a chromatographic column is combined with the determination of the concentration of separated substances using various detectors (spectrophotometric, electrochemical, mass spectrometric, etc.). This is not just a sequence of methods: the characteristics of one initial method, e.g., separation in a column, affect the peculiarities and characteristics of the other, and sometimes dictate them.

In English science literature, in addition to the relatively rarely used term "hybrid methods," is the term "hyphenated methods." Hyphenated methods are widely used in practice, they are used to carry out millions of analyses, including the analyses of complex mixtures of organic substances.

Concentration of trace elements by distilling a matrix, including as a result of chemical reactions, has found many practical applications. This technique is used in the analysis of inorganic substances of high purity (Nizhny Novgorod, Novosibirsk, and Moscow).

6.3 Liquid–Liquid Extraction

In the 1950s–1980s, research and practical developments in the field of liquid–liquid extraction were carried out in the USSR on a very large scale. With the help of this method, important tasks like the radiochemical processing of irradiated

АКАДЕМИЯ НАУК СССР

Н. М. Кузьмин
Ю. А. Золотов

КОНЦЕНТРИРОВАНИЕ
СЛЕДОВ
ЭЛЕМЕНТОВ

Fig. 6.1 Title page of N. M. Kuz'min and Yu. A. Zolotov's book [2] on the preconcentration of trace elements

nuclear fuel, hydrometallurgy (uranium, niobium, tantalum, zirconium, rare-earth elements, etc.) and, of course, analytical chemistry, were solved. In chemical analysis, extraction was widely used for the extraction-photometric (fluorometric) determination of elements, for selective isolation and concentration of elements before their determination by other methods, and for group concentration of microelements prior to their determination, e.g., by atomic-emission or X-ray fluorescence methods. The Russian scientific contribution to this field was significant; apparently, one of the largest.

The theory of chelate extraction has developed, especially in terms of solvent choice for coordinatively saturated and coordinatively unsaturated chelates (donor-active solvents for unsaturated), and additionally in terms of extracting charged hydrophilic chelates into the organic phase [6]. The notion of the hydrate–solvate mechanism of acid extraction was introduced [7, 8] and the phenomenon of suppression of extraction of one element by another was discovered [9] (Fig. 6.2).

Yu. A. Zolotov and B. Ya. Spivakov proposed a theory of exchange extraction of chelates. These same researchers, along with O. M. Petrukhin developed the theory of extraction of metal complexes from the standpoint of coordination chemistry, and contributed to the theory of synergistic extraction. The effects of stability, charge, and hydration of complexes on their extraction were revealed. As a result, recommendations for the selection of optimal extraction systems were formulated.

B. Ya. Spivakov proposed the use of a two-phase aqueous system based on water-soluble polymers for the extraction of metal complexes using water-soluble ligands. The chemistry of metal extraction in polymer–salt–water and polymer–polymer–water systems has been studied [10, 11]. The use of such systems greatly expanded the range of extraction reagents and made it possible to work in the absence of conventional organic solvents, which improved the safety aspects of working with extraction systems. This work was later developed in the United States, where similar systems were used to separate radionuclides. Water-soluble polymers, forming complexes with metals, have also been proposed for use in membrane filtration. This method has found applications in analytical practice and radiochemistry.

This work was performed by Yu. A. Zolotov and co-workers at the V. I. Vernadskii Institute of Geochemistry and Analytical Chemistry of the USSR Academy of Sciences. In 2005, Yu. A. Zolotov was awarded the C. Hanson Medal by the International Committee of Extraction Chemistry and Technology, for his contribution to the development of the extraction method. For his work on extraction he was also awarded a gold D. I. Mendeleev Medal by the Russian Academy of Sciences (RAS) and the D. I. Mendeleev Russian Chemical Society. Yu. A. Zolotov, O. M. Petrukhin, and B. Ya. Spivakov received the L. A. Chugaev Award of the RAS, and Yu. A. Zolotov, B. Ya. Spivakov, and L. N. Moskvin—the V. G. Khlopin Award of the RAS.

V. I. Kuznetsov proposed so-called low-melting extractants (at the time of extraction they are liquid, they solidify upon cooling to room temperature, and are then easily separated) [12]. These are convenient, e.g., for the subsequent X-ray fluorescence determination of elements [13]. A group headed by V. P. Zhivopistsev, B. I. Petrov, and M. I. Degtev, in Perm, developed extraction in three-phase systems (for concentrating substances in the third phase) [14, 15].

New extragents were proposed. Organic sulphides (specialists from Ufa and Novosibirsk), diantipyrilmethane (S. I. Gusev, 1950), and 1-phenyl-3-methyl-4-benzoylpyrazolone-5 (PMBP) (Yu. A. Zolotov and M. K. Chmutova) gained widespread use. A new class of organometallic extractants for anions was proposed and studied, in which the complexing atom is not oxygen, nitrogen, or sulfur, but a

Доклады Академии наук СССР
1968. Том. 180, № 6

УДК 541.123:542.61:543.72 *ХИМИЯ*

Ю. А. ЗОЛОТОВ

СОЭКСТРАКЦИЯ И ПОДАВЛЕНИЕ ЭКСТРАКЦИИ КОМПЛЕКСНЫХ МЕТАЛЛОКИСЛОТ

(Представлено академиком А. П. Виноградовым 6 XII 1967)

В работах Карабаша, Мосеева и др. ([1–4]) наблюдалась соэкстракция микроэлементов с макроэлементами, которые извлекали в виде комплексных металлгалогенидных кислот. Например, In, Sn (IV) или Sb (V) соэкстрагировались с железом (III) при извлечении последнего диэтиловым или диизопропиловым эфирами из солянокислых растворов. Как было показано, соэкстрагируются лишь те элементы, которые сами в какой-то степени извлекаются в данных условиях, т. е. речь идет об увеличении коэффициентов распределения одних экстрагирующих элементов в присутствии других. Соэкстракция наблюдалась при использовании простых эфиров, кроме β, β′-дихлордиэтилового, но отсутствовала при использова-

Fig. 6.2 Fragment of the article by Yu. A. Zolotov on the suppression of extraction of one element by another

metal (tin) atom interacting with the oxygen atoms of the anion. The unique quality of these reagents is that they have a high extraction capacity with respect to anions with high hydration energy; they are still the best known extractants for oxygen-containing anions, primarily arsenate and phosphate ions.

A large series of work was carried out in Voronezh, under the leadership of Ya. I. Korenman, on the extraction of phenols and many other organic compounds. New ionic liquids (and new methods for their use) were applied for extraction by I. V. Pletnev at the M. V. Lomonosov Moscow State University [16]. V. N. Bekhterev (Sochi) developed a method of freezing extraction.

Analysts in the USSR developed a vast number of procedures for determining elements, or their groups, using extraction in a liquid–liquid system. I. P. Alimarin, A. I. Busev, I. A. Blyum, V. P. Zhivopistsev, V. M. Peshkova, Yu. A. Zolotov, B. Ya. Spivakov, O. M. Petrukhin, V. G. Torgov, A. G. Karabash, and many others, are among those who formulated these procedures. Some procedures have been widely used and are used in the practice of analysis, e.g., in the determination of platinum metals or gold. Many books on liquid extraction have been published, e.g. [17, 18].

6.4 New Sorbents and Their Application to Preconcentration

It is believed that T. E. Lovitz (1757–1804) in St. Petersburg discovered, in 1796, adsorption from solutions (on charcoal), and used it for technological and analytical purposes.

To preconcentrate traces elements before their determination by various, mainly spectrometric methods, many complexing sorbents were created and used [19].

Myasoedova and co-workers [20, 21] synthesized a series of selective complexing polymeric sorbents with groups of 3(5)-methylpyrazole, imidazole, benzimidazole, quinolines, 2-mercaptobenzothiazole, amidoxime, thioglykolanilide, and arsenazo, for concentrating noble, rare, and other metals. Promising areas of application in inorganic analysis are: group and selective preconcentration of platinum group elements and gold in their determination in natural and industrial materials, the concentration of gold, silver, uranium, rare-earth elements, and some heavy metals, and their determination in natural and waste water. Using POLIORGS (a trademark), adsorbents combined procedures for determining these elements in various objects were developed, including sorption concentration and subsequent determination by atomic-emission, such as ICP, atomic absorption, X-ray fluorescence, neutron activation, spectrophotometry, and other methods. These sorbents are also promising for the recovery of valuable metals from solutions, marine and waste water management, sewage treatment, and the extraction of platinum group metals from process solutions. Selective sorbents of a new type—fibrous materials filled with complexing ion exchangers—were proposed and used for the concentration of elements.

In 1984 a monograph entitled *Chelate-Forming Sorbents* was written, by Myasoedova and Savvin [22], in which the syntheses of complexing sorbents based on various polymeric matrices were presented, physicochemical and analytical properties of the sorbents described, and areas of application characterized (Fig. 6.3).

Fig. 6.3 Galina Vladimirovna Myasoedova developed a series of sorbents called "Polyorgs," along with analysis methods based on their use. Photo provided by B. F. Myasoedov

Galina Vladimirovna Myasoedova *was born on July 20, 1929. She graduated from the D. I. Mendeleev Moscow Institute of Chemical Technology (1953). She is a doctor of chemical sciences and a Leading Researcher of the V. I. Vernadskii Institute of Geochemistry and Analytical Chemistry of the RAS.*

One group of sorbents, developed by G. I. Tsyzin and A. A. Formanovskii, was based on the inoculation of conformationally mobile complex-forming groups to the polymer matrix [23, 24]. The sorbents were designed to simultaneously concentrate a large number of elements before being determined by multi-element spectroscopic methods. DETATA sorbents (a diethylenetriamine tetraacetate group based on, e.g., a polystyrene or cellulose matrix), in almost all respects, exceeded the benefits of other commercially available sorbents—Chelex 100, etc. (Fig. 6.4).

With the use of this sorbents in the form of filters, a procedure widely used (in hundreds of laboratories) for determining a large number of elements by the X-ray fluorescence method has been developed. Fluorine-containing sorbents have been proposed as a means of concentrating hydrophobic organic compounds [25]. A methodology of dynamic sorption concentration, used to solve many practical problems, has been developed [26] (Figs. 6.5 and 6.6).

Grigorii Il'ich Tsyzin *was born on March 31, 1957. He is a doctor of chemical sciences, a professor, and a Principle Researcher of the Division of Analytical Chemistry of the Chemistry Department of the M. V. Lomonosov Moscow State University. His work is in the field of analytical preconcentration of elements and organic compounds. He is laureate of the V. A. Koptyug Award of the RAS.*

The series of V. N. Losev's work deals with the use of chemically modified silica for sorption concentration, separation, and subsequent spectroscopic determination of noble and non-ferrous metals [27, 28]. Research is being conducted on the synthesis of sorbents based on vegetable materials, chemically modified with various functional groups, and based on inorganic oxides, modified with polymeric polyamines and sulfo derivatives of organic reagents. These sorbents are used in sorption spectroscopic and testing methods for determining metal ions in objects of various compositions.

Fig. 6.4 Structure of DETATA sorbents

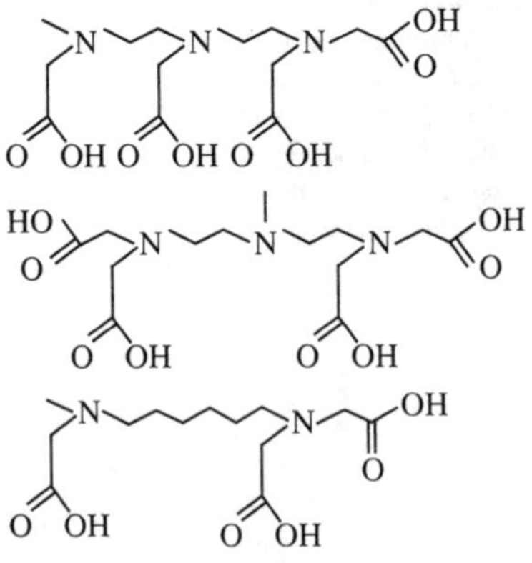

Fig. 6.5 Grigorii Il'ich Tsizin developed effective "DETATA" sorbents and methods linked to their applications. Photo provided by G. I. Tsysin

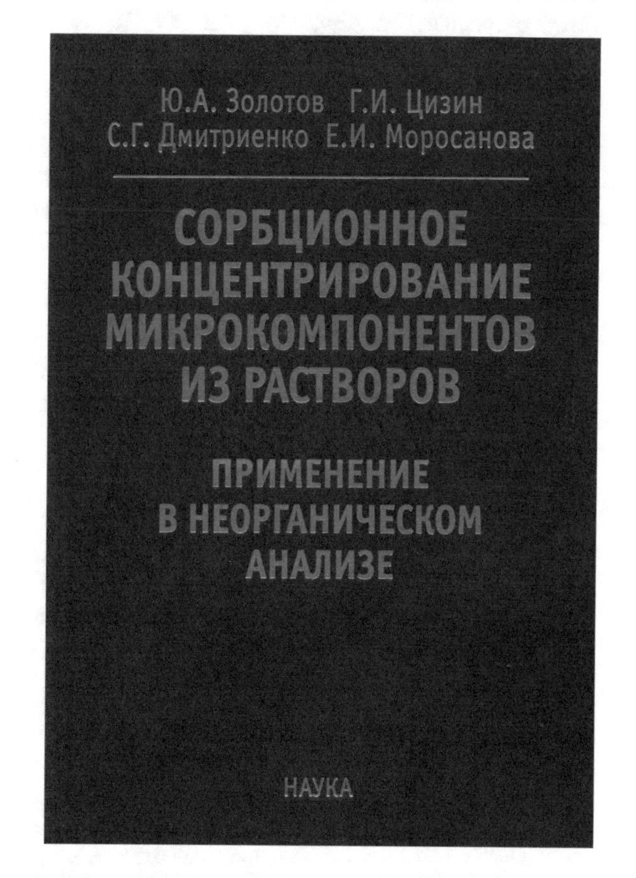

Fig. 6.6 Books about sorption preconcentration [19]

In recent years, work has been carried out to determine the elements and their speciation in biological objects. Much work has been done in this direction, including work of fundamental character [29, 30].

A vast amount of work on the use of polyurethane foam, especially in modified form, as a sorbent and analytical reagent has been carried out by S. G. Dmitrienko and co-workers [31]. Super-cross-linked polystyrene, created by V. A. Davankov, turned out to be a very successful sorbent for concentrating microcomponents; it is currently widely used [32].

6.5 Capillary Isotachophoresis and Other Separation Techniques

The method of isotachophoresis was developed in 1923 by J. Kendall and E. Crittenden. Capillary isotachophoresis was proposed in the Leningrad Physico-Technical Institute of the USSR Academy of Sciences by B. P. Konstantinov and O. V. Oshurkova [33] (Fig. 6.7).

A device for capillary isotachophoresis was made in the 1950s. In 1957–1960 it was shown at exhibitions in the USSR, Poland, and Bulgaria. A Swedish firm (apparently LKB) wanted to purchase a license for the device, but the developers could not agree with the proposal and preferred the scientific publication route. The method was subsequently developed in Czechoslovakia and other countries; it was improved by Evaraerts [34, 35]. At a later time, a few large companies—LKB (Sweden) and Shimadzu Seisakusho (Japan)—began to produce instruments for capillary isotachophoresis.

V. I. Kuznetsov proposed a new method for concentrating elements—coprecipitation with organic coprecipitates [36, 37]. In the 1950s, the sensitivity of methods for determining the elements was insufficient, and the need to determine

Доклады Академии наук СССР
1963. Том 148, № 5

ФИЗИЧЕСКАЯ ХИМИЯ

Академик Б. П. КОНСТАНТИНОВ, О. В. ОШУРКОВА

ЭКСПРЕССНЫЙ МИКРОАНАЛИЗ ХИМИЧЕСКИХ ЭЛЕМЕНТОВ МЕТОДОМ ДВИЖУЩЕЙСЯ ГРАНИЦЫ

В аналитической химии существует много различных направлений для разработки количественных и качественных методов анализа Настоящей работой открывается одно из новых направлений в этой области — экспрессный микроанализ методом движущейся границы. Метод движущейся границы известен давно, со времени работы Лоджа ([1]), который пытался использовать его для определения абсолютных скоростей движения ионов. Впоследствии этот метод применялся для определения чисел переноса ионов ([2,3]) для разделения химических элементов ([4,5]), для разделения изотопов ([6,7]) Лонгсворт ([8]) попытался использовать этот метод для количест-

Fig. 6.7 Fragment of the first article on capillary isotachophoresis

trace amounts, especially of rare and scattered elements, increased. Organic coprecipitators made it possible to solve a number of complex analytical problems. Under the guidance of V. I. Kuznetsov this approach was developed theoretically, and procedures for concentrating almost all the elements, including actinides, from very dilute solutions, were developed.

A. B. Blank in Kharkov (Institute of Single Crystals) developed one more method of preconcentration—directed crystallization [38]. The method was used in the analysis of a number of high-purity substances.

References

1. Zolotov, Y.A., Kuz'min N.M.: Preconcentration of Trace Elements, 288 pp. Khimiya, Moscow (1982) (in Russian)
2. Kuz'min, N.M., Zolotov, Y.A.: Preconcentration of Trace Elements, 268 pp. Nauka, Moscow (1988) (in Russian)
3. Zolotov, Y.A., Kuz'min, N.M.: Preconcentration of Trace Elements (Comprehensive Analytical Chemistry, 25), 372 pp. Elsevier, Amsterdam (1990)
4. Zolotov, Y.A.: Zh. Anal. Khim. **32**, 2085 (1977)
5. Zolotov, Y.A.: Analyst **103**, 56 (1978)
6. Zolotov, Y.A.: Extraction of Chelate Compounds, 310 pp. Ann Arbor—Humphrey Science Publishers, London (1970)
7. Zolotov, Y.A.: Dokl. AN. SSSR. **145**, 100 (1962)
8. Zolotov, Y.A.: Usp. Khim. **32**, 220 (1963)
9. Zolotov, Y.A.: Dokl. AN. SSSR. **180**, 1367 (1968)
10. Zvarova, T.I., Shkinev, V.M., Spivakov, B.Ya., Zolotov, Y.A.: Dokl. AN. SSSR, **107**, 273 (1983)
11. Zvarova (Nifant'eva), T.I., Shkinev, V.M., Spivakov, B.Ya., Zolotov, Y.A.: Microchim. Acta. **III,** 449 (1984)
12. Kuznetsov, V.I., Seryakova, I.V.: Zh. Anal. Khim. **14**, 161 (1959)
13. Lobanov, F.I.: Extraction of inorganic compounds by melts of organic substances. In: The Results of Science and Technology, Ser. Inorganic Chemistry, 7, 84 pp (1980) (in Russian)
14. Zhivopistsev, V.P., Mochalov, K.I., Petrov, B.I., Yakovleva, T.P.: To the formation of three-phase systems in the extraction of elements by diantipyryl methane. In: Chemistry of Extraction Processes, p. 194. Nauka, Moscow (1972) (in Russian)
15. Petrov, B.I., Kalitkin, K.V., Nazemtseva, K.A.: Izv. Altaisk. Gos. Un-ta. **3**(79), 198 (2013)
16. Pletnev, I.V., Smirnova, S.V., Khachatryan, K.S., Zernov, V.V.: Ross. Khim. Zh. (Zh. Ross. Khim. Ob.), **48**, 51 (2004)
17. Blum, I.A.: Extraction-Photometric Methods Using Basic Dyes, 220 pp. Nauka, Moscow (1970) (in Russian)
18. Zolotov, Y.A.: Extraction in Inorganic Analysis, 83 pp. Moscow State University, Moscow (1988) (in Russian)
19. Zolotov, Y.A., Tsyzin, G.I., Dmitrienko, S.G., Morosanova, E.I.: Sorption Preconcentration of Microcomponents from Solutions. Application in Inorganic Analysis, 320 pp. Nauka, Moscow (2007) (in Russian)
20. Myasoedova, G.V., Savvin, S.B.: Crit. Rev. Anal. Chem. **17**, 1 (1986)
21. Myasoedova, G.V.: Solv. Extr. Ion. Exch. **6**, 301 (1988)
22. Myasoedova, G.V., Savvin, S.B.: Chelate-Forming Sorbents, 174 pp. Nauka, Moscow (1984) (in Russian)

23. Tsyzin, G.I., Seregina, I.F., Sorokina, N.M., Formanovskii, A.A., Zolotov, Y.A.: Zavodsk. Lab. Diagn. Mater. **59**, 3 (1993)
24. Varshal, G.M., Velyukhanova, T.K., Pavlutzkaya, V.I., Starshinova, N.P., Formanovsky, A. A., Seregina, I.F., Shilnikov, A.M., Tsysin, G.I., Zolotov, Y.A.: Int. J. Environ. Anal. Chem. **57**, 107 (1994)
25. Arkhipova, A.A., Statkus, M.A., Tsyzin, G.I., Zolotov, Yu.A.: Zh. Anal. Khim. **70**, 1235 (2015)
26. Tsyzin, G.I., Statkus, M.A. Sorption Concentration of Microcomponents Under Dynamic Conditions, 480 pp. LENAND, Moscow (2016) (in Russian)
27. Losev, V.N., Borodina, E.V., Buiko, O.V., Maznyak, N.V., Trofimchuk, A.K.: Zh. Anal. Khim. **69**, 462 (2014)
28. Didukh, S.L., Losev, V.N., Mukhina, L.N., Trofimchuk, A.K.: Zh. Anal. Khim. **71**, 1137 (2016)
29. Lisichkin, G.V. (ed.).: Modified Silica in Sorption, Catalysis, and Chromatography, 248 pp. Khimiya, Moscow (1986) (in Russian)
30. Kholin, Y.V.: Quantitative Physicochemical Analysis of Complexation in Solutions and on the Surface of Chemically Modified Silica: Content Models, Mathematical Methods, and Their Applications, 288 pp. Folio, Khar'kov (2000) (in Russian)
31. Dmitrienko, S.G., Apyari V.V.: Polyurethane Foams: Sorption Properties and Application in Chemical Analysis, 264 pp. KRASAND, Moscow (2010) (in Russian)
32. Davankov, V.A., Tsyurupa, M.P.: HyperCrosslinked Polymeric Networks and Adsorbing Materials, Synthesis, Structure, Properties and Applications, 648 pp. Elsevier, Amsterdam (2010)
33. Konstantinov, B.P., Oshurkova, O.V.: Dokl. AN. SSSR. **148**, 1110 (1963)
34. Evaraets, F.M., Verheggen, T.P.E.M.: J. Chromatogr. **73**, 193 (1972)
35. Martin, A.J.P., Evaraerts, F.M.: Proc. Royal Soc. London **A316**, 493 (1970)
36. Kuznetsov, V.I.: Zh. Anal. Khim. **4**, 9 (1954)
37. Kuznetsov, V.I., Akimova, T.G.: Preconcentration of Actinides by Coprecipitation with Organic Coprecipitators, 277 pp. Atomizdat, Moscow (1968) (in Russian)
38. Blank, A.B.: Analysis of Pure Substances Using Crystallization Preconcentration, 184 pp. Khimiya, Moscow (1986) (in Russian)

Chapter 7
Other Methods

Abstract Several very original methods of analysis have been developed. Among them is the so-called stoichiographic method of differentiating dissolution, which is a new method for the phase analysis of complex solid objects (V. V. Malakhov). An exceptionally sensitive, but complicated, method of molecular condensation nuclei was proposed by Ya. I. Kogan. In Russia, there has been a drive toward test methods, especially for "field" analysis. L. A. Gribov and M. E. Elyashberg developed a system for elucidation of the structures of organic compounds based on spectral data (expert systems). There are indications that the first semiconductor gas sensors were created by I. A. Myasnikov and others.

7.1 General Remarks

This chapter includes very different methods developed in Russia, which, because of their original nature, do not fall into any of the groups already considered. Some of these methods are recognized by specialists and are used practically across the globe; others, including very curious methods, for various reasons have not yet been widely used (and one, perhaps, will not be). Acquainting myself with some of these methods, I feel the urge to express my admiration for the ingenuity and creativity of their developers.

What are these methods? Here we will simply name them and consider their characteristics briefly; later in this chapter they are discussed in more detail.

One such method is the so-called stoichiographic method of differentiating dissolution, developed by V. V. Malakhov. This method is for the phase analysis of inorganic substances and materials that are complex in chemical and mineralogical composition (multiphase). It consists of carrying out a continuous differentiating dissolution of the object under analysis, in a dynamic manner, and a continuous multi-element analysis of the resulting solution. This method is implemented in a special piece of apparatus called a stoichiograph.

The second method, with a lesser-known name—the method of molecular condensation nuclei (MCN)—was proposed by Ya. I. Kogan. This method is based

Y. A. Zolotov, *Russian Contributions to Analytical Chemistry*,
https://doi.org/10.1007/978-3-319-98791-0_7

on the fact that under certain conditions all molecules of organic matter present in vapor and gas become centers of condensation. Roughly speaking, each molecule results in the formation of one drop. Drop concentration may be determined by nephelometry (turbodimetry). The sensitivity of this method is very high.

In this chapter, test determinations are also discussed—chemical and biochemical. In general, this is certainly not a Russian invention, but useful steps have been taken in Russia in this direction.

L. A. Gribov, M. E. Elyashberg and co-workers developed their system for elucidation of the structure of molecules of organic compounds based on spectroscopic data—nuclear magnetic resonance (NMR) and vibrational spectroscopy. Approaches based on expert systems (artificial intelligence systems) and deep understanding of the physics of molecules, are used.

Finally, there is a small section on semiconductor gas sensors in this chapter, the concept of which seems to be connected with the name I. A. Myasnikov, an employee of the Ya. L. Karpov Moscow Physico-Chemical Institute.

7.2 Stoichiographic Method of Differentiating Dissolution

In 2009, the founder of this method of phase analysis, V. V. Malakhov (Fig. 7.1), wrote approximately the following:

> For many years in the journals on analytical chemistry there are few articles on phase analysis. The problem appeared to be too complicated, e.g., for the most well-known chemical method of phase analysis – selective dissolution. How to separate a mixture of solid phases and find out what the phases are, if only it is known that in the mixture, e.g., there are ten elements. Until recently, in analytical chemistry there were no methods suitable for the effective solution of such problems. The actual monopoly of physical methods resulted in rejection of the 'wet' chemical methods of phase analysis. The complexity of the problem of separation of solid phases resulted in a loss of interest and the analysts themselves in solving the problems of determination in solid multiphase mixtures of both stoichiometric composition, and the content of each phase. But we are not talking about any particular issues and tasks. Phase analysis is a huge area of inorganic analysis that is huge in scope and in variety of objects.

The situation changed after 1986, when the stoichiographic method of differentiating dissolution was developed [1, 2].

The essence of the method lies in the fact that a solid multicomponent, multiphase system (e.g., a complex catalyst or a rock) is subjected to gradient dissolution. The composition of the solvent, e.g., an acid solution, changes during dissolution in terms of increasing its dissolving power. As a result, various components of the complex system are successively dissolved. The solution obtained after dissolution starts is continuously analyzed (usually by atomic emission spectroscopy with inductively coupled plasma). The device used to record this monitors not only the concentration of the solution's chemical elements, but also their ratios. It is the ratios which make it possible to conclude which compounds (phases) are dissolved at any given moment (Fig. 7.2).

Fig. 7.1 Vladislav Veniaminovich Malakhov developed the stoichiographic method of differentiating dissolution. Photo provided by V. V. Malakhov

Fig. 7.2 Malakhov employee A. A. Pochtar' at work on a stoichiograph. Photo provided by V.V. Malakhov

The device for flow analysis (stoichiograph) was created at the G. K. Boreskov Institute of Catalysis of the Siberian Branch of the Russian Academy of Sciences (RAS). The stoichiograph includes tanks for solvent components, peristaltic pumps with an electronic system for programming the composition and temperature of the solvent, a mixer for the solvent components, a thermostatted flow reactor, a sampling and sample preparation unit, an element composition detector, and an information storage and processing unit. Separate blocks of the complex are connected by a system of capillaries. As previously reported, the ICP AES spectrometer is used as a detector, which allows the determination in the flow of the resulting solution simultaneously at intervals of 11 s for 38 elements at a time (plus one of the others, optionally). It has detection limits of 10^{-2}–10^{-3} μg/mL and an error of 1–5% rel. In the course of dissolution, several thousands of measurements can be made, which makes the recording of the kinetics appear virtually continuous.

With the help of this method the deciphering of the phase composition of various objects has been solved. For more details about this the method, see, e.g. [3–5].

Vladislav Veniaminovich Malakhov *was born on October 20, 1934. He is a doctor of chemical sciences, a professor, and Principle Researcher at the G. K. Boreskov Institute of Catalysis of the Siberian Branch of the RAS. He graduated from the S. M. Kirov Kazakh University in Alma-Ata (1957), worked as an engineer in the laboratory of the Leninogorsk Polymetallic Plant. He attended a graduate course at the Kazakh University and defended his candidate of sciences thesis. In 1964, he joined the Institute of Inorganic Chemistry of the Siberian branch of the RAS in Novosibirsk, and in 1966 he headed the analytical laboratory of the Institute of Catalysis. He led research on the development of polycapillary chromatography (PCC), and in 1979 the first column (PCC) was manufactured. Based on PCC, express analyzers of the "Ekho" type were assembled. V. V. Malakhov solved problems of analysis of substances of unknown phase composition. In 1988 he defended his doctoral dissertation entitled "Chemical methods of phase analysis of heterogeneous catalysts." He authored about 250 scientific publications, and originated several patents. He was chairman of the Siberian Branch of the Scientific Council of the RAS in Analytical Chemistry. He repeatedly chaired the organizing committees of the Analytics of Siberia and the Far East conferences.*

7.3 Method of Molecular Condensation Nuclei

In the early 1960s, Ya. I. Kogan developed a new method for determining impurities in gases. Each molecule of an impurity, under certain conditions, becomes the nucleus of an aerosol particle, the concentration of such particles is determined nephelometrically. The method appeared to be one of the most sensitive in modern analytical chemistry.

Aerosol particles are formed from individual molecules of determined composition (or, more often, the product of its transformation) in a supersaturated vapor of specially introduced volatile organic substances with high molecular weights. The operations which provide the formation of aerosol particles from the impure compound's molecules are carried out in a continuous flow of analyzed gas or analyzed gas mixed with carrier gas.

The molecules of most compounds present as impurities in a gas are not themselves capable of serving as condensation nuclei. Therefore, they are activated by exposure to ultraviolet light, electric discharge, high temperature, or subjected to chemical transformation. With such activation, compounds capable of acting as condensation nuclei are formed. These are, e.g., metal oxides, silicon dioxide, and coordinatively unsaturated compounds of $SnCl_4$ type. Thus, in the first work of Ya. I. Kogan it was necessary to determine the ultra-low concentrations of nickel carbonyl. This carbonyl, by the action of oxygen at 60 °C, was converted into NiO_2 molecules which can act as condensation nuclei.

The second (not mandatory) operation is the activation of molecular condensation nuclei (MCN). This operation is resorted to if one wants to achieve the highest sensitivity of determinations. The flow of the analyzed gas is acted upon by the vapors of an activating substance, e.g., oxalic acid. Molecules of activators join the already formed MCN and increase the probability of conversion of nuclei into aerosol particles by several orders of magnitude.

The third operation is called manifestation. Condensation nuclei are affected by the developing vapor which is a non-volatile organic substance capable of interacting with condensation nuclei—irreversibly growing aerosol particles. Their enlargement is facilitated by another stage—treatment with supersaturated vapor of diisobutyl phthalate; monodisperse particles with diameters of ~ 0.5 μm are formed, the concentration of which are conveniently measured by a nephelometer, a process which is carried out in the final stage.

Aerosol gas analyzers have existed before; however, the difference with the MCN method is that aerosol particles are formed on each molecule of an impurity, while in other aerosol analyzers the aerosol particle is formed as a result of coagulation of a large number of molecules. Hence, the extremely low detection limit achieved using the MCN method. In addition, the method is characterized by a large linear range of analytical signals.

On the basis of the principle considered, MCN detectors, suitable as chromatographic detectors, have been developed. In this case, a large consumption of carrier gas is required, but there are ways in which this can be reduced; in addition, dry, clean air may be used as the carrier gas.

Examples of achievable detection limits of impurities in gases using the MCN method are (in mg/L): carbonyls of metals – up to 10^{-13}, halides of the elements of groups III–IV of the periodic table—up to 10^{-10} (a flame photometric detector provides a detection limit of 10^{-4}), organophosphorus compounds—10^{-8}, sulfur dioxide—10^{-8} (FPD—3×10^{-6}), and ammonia—3×10^{-6}.

The method was developed at the Neorganika Co. (Elektrostal, Moscow region), where devices have also been developed for its application in solving various analytical problems. The method, and corresponding instruments, were first used to determine metal carbonyls. Subsequently, the method was applied to control the efficiency of filters for cleaning gases, including air, and in checking the tightness of critical products—heat exchangers for fast neutron reactors, fuel elements for nuclear power plants, caisson tanks in aircraft wings, etc.

An important area for the use of the MCN method is the determination of chemicals present when chemical weapons are destroyed. This is because the method offers sufficient sensitivity for monitoring not only the working area, but also the surrounding atmosphere. Thus, the limit of detection of lewisite is $\leq 10^{-8}$ g/m^3 which has a maximum permissible concentration (MPC) within the work area of 2×10^{-7}. Examples of other detectable substances are: 1,1-dimethylhydrazine, which is a liquid propellant component for which the detection limit is up to 10^{-7} g/m^3 at an MPC within the working area of 1×10^{-4}; and nickel tetracarbonyl with a detection limit of 10^{-12} g/m^3.

The first open publication on the MCN method dates back to 1965 [6].[1] Subsequently, Ya. I. Kogan published a fairly large number of articles; all of them are quoted in [7]; he also authored a small book [8] (Fig. 7.3).

Fig. 7.3 Ya. I. Kogan's book about the molecular nuclei of condensation

Министерство образования Российской Федерации

САНКТ-ПЕТЕРБУРГСКИЙ
ГОСУДАРСТВЕННЫЙ ПОЛИТЕХНИЧЕСКИЙ УНИВЕРСИТЕТ

Я.И. КОГАН

МОЛЕКУЛЯРНЫЕ ЯДРА КОНДЕНСАЦИИ И СОПУТСТВУЮЩИЕ ЯВЛЕНИЯ

САНКТ-ПЕТЕРБУРГ
ИЗДАТЕЛЬСТВО СПБГПУ
2003

[1]As far as I remember, I helped publish this article—*Yu.Z.*

7.4 Express Tests Methods

It is almost impossible to draw a clear line between "normal" methods and test methods of analysis. Moreover, we observe how "conventional" methods, being embodied in clever techniques and easy-to-use inexpensive portable devices, become test methods. Glucometers or nitratomers, which practically became domestic, serve as examples.

Characteristic features which test methods require are: the possibility of mass production, the relative cheapness and ease of use, the absence of the requirement for special education and training to be given to the user, and the ability to directly obtain results in a form understandable to the user. Such methods and means for implementation have long been known and are widely used and produced by many firms. The means of this type is diverse: paper strips, ampoules, indicator tubes, tablets, pocket sets of reagents with plastic test tubes, etc. Such means are convenient, in particular, for school chemistry classrooms, for chemical and

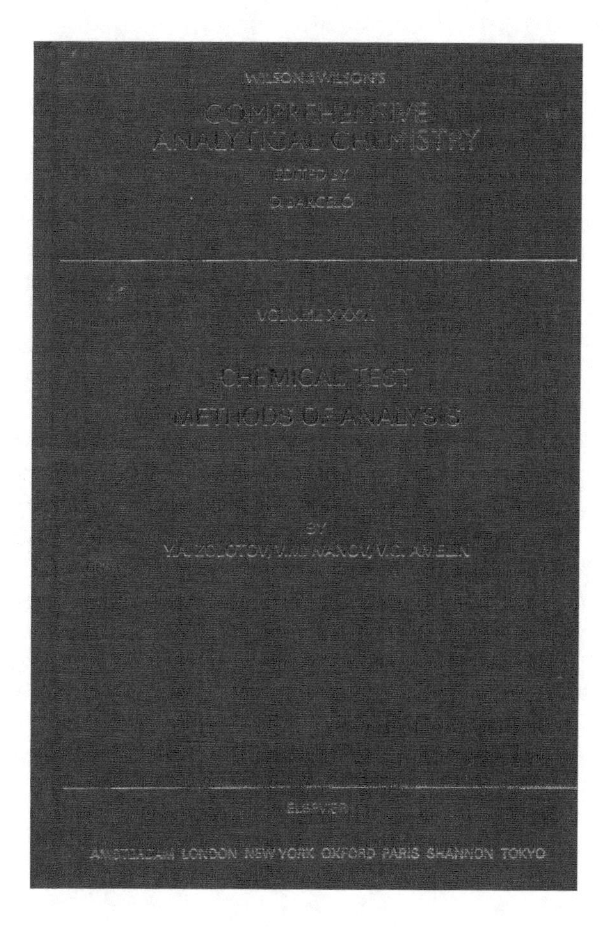

Fig. 7.4 A book on test methods of analysis (2003)

environmental clubs, and expeditions. They are also used by professionals—chemists, geologists, physicians, agronomists, etc.

N. A. Tananaev, at the Kiev Polytechnic Institute in the 1920s, developed, in parallel with the Austrian chemist F. Feigl, so-called drop analysis, oriented, however, more toward use by analysts in a laboratory. The surge of interest in chemical test methods in Russia dates back to the 1990s; it was largely stimulated by the work performed at the Moscow State University, and dealt with the development of field analysis. At this university, numerous test methods and test tools were created (Yu. A. Zolotov, E. I. Morosanova, S. G. Dmitrienko et al.). Among these are: indicator tubes for analysis of solutions; tablets of polyurethane foam with impregnated reagents; and sets for the implementation of chemical and enzymatic reactions. Several small firms were established to produce these test sets. The result of the initial stage of this cycle of work was reflected in a monograph published in Russian and English [9, 10] (Fig. 7.4).

A separate area of development has been in immunoassays, e.g., at the A.N. Bach Institute of Biochemistry of the RAS and at M.V. Lomonosov MSU (B.B. Dzantiev, S.A. Eremin et al.) [11].

7.5 Elucidation of the Structure of Organic Compounds Using Spectroscopic Data

L. A. Gribov, M. E. Elyashberg et al. developed a system for the elucidation of the structures of organic compounds based on spectroscopic data—first data from vibrational spectroscopy, then, increasingly NMR data.

Artificial intelligence (an expert system) was used in this method [12]. Algorithms and programs were created which became commercial products. The ACD/Structure Educator system was acquired by several organizations in Russia (Figs. 7.5 and 7.6).

The use of an expert system makes it possible to detect and correct erroneous structures, which prevents the acquisition of incorrect structures. Today there are examples of the application of this system to identify structures, which by traditional methods of analysis of two-dimensional NMR spectra were considered impossible. The use of atomic force microscopy in combination with an expert system, and the quantum chemical calculation of the geometry and chemical shifts of ^{13}C have made it possible to determine the structure of molecules in cases when X-ray diffraction analysis is impossible. A synergistic combination of empirical and quantum methods for calculating NMR spectra has proved to be an optimal way of distinguishing very close structures (see [13–15]).

Fig. 7.5 Lev Aleksandrovich Gribov (born May 23, 1933) developed expert systems for elusidation of the structures of organic compounds. Photo provided by L. A. Gribov

Fig. 7.6 Mikhail Evgen'evich Elyashberg (born May 7, 1936) developed an effective expert system for evaluating the structures of organic compounds based on nuclear magnetic resonance spectra. Photo provided by M. E. Elyashberg

7.6 Semiconductor Gas Sensors

The principle of operation of semiconductor gas sensors is the change in the electrical characteristics of certain metal oxides when certain gases are sorbed on their surfaces. The magnitude of the electrical signal corresponds to the number of molecules sorbed from the environment or appearing on the surface of the sensor due to a heterogeneous chemical reaction. Although the change in the electrophysical characteristics of oxides upon gas absorption was previously known, the idea of using this effect to determine the gas concentration was simultaneously expressed by Heiland [16] and Myasnikov [17]. The prehistory of the use of semiconductor sensors in chemical analysis (or rather, the history of the development of semiconductor sensors themselves) was briefly described in a book by Myasnikov et al. [18].

"The need to use electronic representations to solve a number of physicochemical problems arising in the investigation of heterogeneous processes was realized by Pisarzhevskii already in the early 1920s [19], and some non-trivial ideas on the effect of adsorption on the electrophysical properties of semiconductor adsorbents were laid down in classical works of Ioffe [20], Roginskii [21] and other researchers. Further theoretical development of these ideas was obtained in the works of Vol'kenshtein and his trainees (see the monograph [22] and the references contained there), as well as in the papers of Hauffe [23] and a number of others.[2]

The main idea of F. F. Vol'kenshtein, the founder of the electronic theory of chemisorption, was that the chemisorbed particle and solid represent a single quantum-mechanical system, in analyzing which it is necessary to take into account changes in the electronic state of both the adparticle and the adsorbent itself [22]. In other words, adsorption in this case is a chemical combination of molecules with an adsorbent."

And now let us quote an excerpt from the same book (p. 23), referring to the actual chemical (gas) sensors.

"The idea, … which consists in the possibility of controlling the composition of the gas atmosphere surrounding the semiconductor, based on an estimation of the changes in its electrophysical characteristics, to the best of our knowledge, was first and practically simultaneously expressed by Heiland [16] and Myasnikov [17]. It is this idea that was the basis of the now widely used method of semiconductor sensors.

The basic premise of the proposal expressed in these papers was the following: a detailed study of the changes in the electrophysical characteristics of a semiconductor adsorbent, caused by the adsorption of a specific gas in a rather wide range of pressures of the latter, makes it possible to solve the inverse problem of determining the concentration of this gas in the volume surrounding the adsorbent due to the change in electrophysical characteristics of the adsorbent caused by its presence."

Many people involved in working with gas sensors consider the Japanese group, who published their work in 1962, i.e., later than Hyland and Myasnikov, to be the originators of this field.

It remains to add that this area of study has received significant development in the form of scientific research and published books, with firms now currently producing a variety of gas sensors.

---

[2]References to other Hauffe publications, and the articles of "a number of others," have been omitted—*Yu.Z.*

7.7 Some Original Novel Devices and Appliances

In addition to the instruments mentioned in this and other chapters,[3] it is possible to name other devices and laboratory installations which were either the first to be developed or have been significantly improved.

In the 1970s, at the V. I. Vernadskii Institute of Geochemistry and Analytical Chemistry of the USSR Academy of Sciences, an atomic absorption spectrometer with a continuous spectrum source (a xenon lamp) was developed. It was a valid and registered model of the device, which, however, could not be put into the series. After many years, similar instruments started to be produced by the firm AnalytikJena.

The first portable ion chromatograph, CPI-1, was assembled in the early 1980s by "Khimavtomatika" company.

Under the leadership of N. S. Grishin, in Kazan, a whole series of devices for sample preparation were developed.

Some instruments produced by the St. Petersburg "Lumex" company are widespread—a mercury analyzer (Fig. 7.7), a device for luminescent analysis known as "Fluorat," and others.

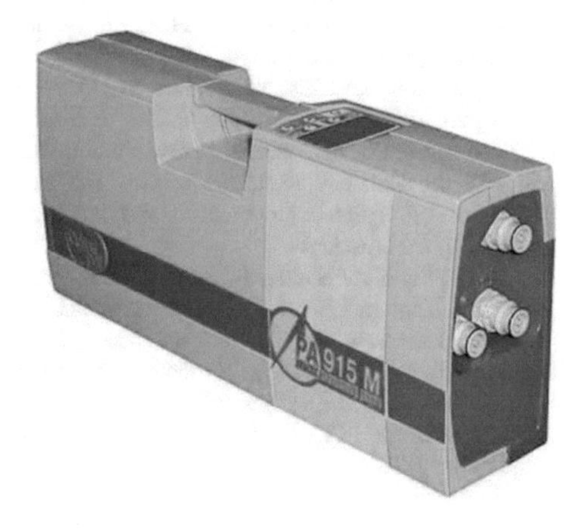

Fig. 7.7 The "Lumex" mercury analyzer was sold in several countries. Photo taken from the advertising brochure of Lumex Co.

[3]For example, the first microcolumn liquid chromatograph ("Ob'", "Milikhrom") was discussed in Chap. 3, numerous devices for mass spectrometry—also in Chap. 2, and the precision coulometer —in Chap. 4.

References

1. Malakhov, V.V.: Dokl. AN SSSR **290**, 1152 (1986)
2. Malakhov, V.V.: Zh. Anal. Khim. **64**, 1125 (2009)
3. Malakhov, V.V.: Zh. Anal. Khim. **44**, 1177 (1989)
4. Malakhov, V.V., Vlasov, A.L., Boldyreva, N.N., Dovlitova, L.S.: Zavodsk. Lab. **62**, 1 (1996)
5. Malakhov, V.V., Vlasov, A.A.: Kinet Catal **36**, 503 (1995)
6. Kogan, Y.I.: Dokl. AN. SSSR. **161**, 388 (1965)
7. Kyandzhetsian, R.A., Katelevskii, V.Y., Valyukhov, V.P., Demin, S.V., Kapashin, V.P., Polkov, A.B., Maiorov, A.V.: Ross. Khim. Zh. (Zh. Ross. Khim. Ob.) **46**, 20 (2003)
8. Kogan, Y.I.: Molecular Condensation Nuclei and Accompanying Phenomena, 75 pp. Izd. SPbGPU, St.Petersburg (2003) (in Russian)
9. Zolotov, Y.A., Ivanov, V.M., Amelin, V.G.: Chemical Test Methods of Analysis, 304 pp. Editorial URSS, Moscow (2002) (in Russian)
10. Zolotov, Yu.A., Ivanov, V.M., Amelin, V.G.: Chemical Test Methods of Analysis (Comprehensive Analytical Chemistry, 36), 317 pp. Elsevier, Amsterdam (2002)
11. Dzantiev, B.B.: Immunoanalytical Methods. In: Dzantiev, B.B (ed.) Biochemical Methods of Analysis (Problems of Analytical Chemistry, 12), 303 pp. Nauka, Moscow (2010) (in Russian)
12. Gribov, L.A., Elyashberg, M.E.: Crit. Rev. Anal. Chem. **8**, 111 (1979)
13. Elyashberg, M.E., Williams A.J.: Computer-based Structure Elucidation from Spectral Data. The Art of Solving Problems, 447 pp. Springer, Heidelberg (2015)
14. Elyashberg, M.E.: Trends in Anal. Chem., **69**(spec. issue), 88 (2015)
15. Elyashberg, M.E., Williams, A.J., Blinov, K.A.: Contemporary Computer Assisted Approaches to Molecular Structure Elucidation, p. 468. RSC Publishing, Cambridge (2012)
16. Heiland, G.: Zeitsch. Phys. **148**, 15 (1957)
17. Myasnikov, I.A.: Zh. Anal Khim. **31**, 1721 (1957)
18. Myasnikov, I.A., Sukharev, V.Y., Kupriyanov, L.Y., Zav'yalov, S.A.: Semiconductor Sensors in Physical and Chemical Studies, 327 pp. Nauka, Moscow (1991) (in Russian)
19. Pisarzhevskii, L.V.: Selected Works, 273 pp. Izd. AN USSR, Kiev (1936) (in Russian)
20. Ioffe, A.F.: Reports on Scientific and Technical Works in the Republic: Catalysis, 53. NTKhI (1930) (in Russian)
21. Roginskii, S.Z., Shultz, E.I.: Ukr. Khim. Zh. **3**, 177 (1928)
22. Vol'kenshtein, F.F.: Electronic Processes on the Surface of Semiconductors During Chemisorption, 345 pp. Nauka, Moscow (1987) (in Russian)
23. Hauffe, K.: Reactions in Solids and on Their Surfaces, vol. 1, 456 pp. Izd-vo inostr. lit., Moscow (1963) (in Russian)
24. Beloglazova, N.V., Eremin, S.A.: Anal. Bioanal. Chem. **407**, 8525 (2015)

Chapter 8
Analysis of Specific Objects. Analytical Chemistry of Individual Analytes and Their Groups

Abstract In Russian laboratories, many problems related to analyzing nuclear materials have been solved; in this area, a great deal of experience has been accumulated. The same can be said about the analysis of materials for microelectronics and other high-purity substances. A significant contribution to the solution of practical problems was made by A. P. Vinogradov, I. P. Alimarin, B. F. Myasoedov, Yu. A. Karpov, I. G. Yudelevich, and others. Development of standard reference. Materials has been undertaken.

8.1 General Remarks

A very significant part of the effort made by research analysts in Russia, and around the world, has been spent on solving problems linked to the title of this chapter. Indeed, most of the publications in scientific journals have dealt with new methods of identifying, or especially quantifying, certain substances in specific objects under analysis.

The focus of Russian researchers, on groups of objects to be analyzed, has shifted in accordance with the following hierarchy: mineral raw materials; metals and alloys; atomic materials; various substances of high purity, especially for electronics; environmental objects; and medicinal substances, food products, and bio-objects.

The greatest contributions of Russian scientists have been made to the analyses of mineral raw materials, especially to the determination of platinum metals and other rare elements; of high-purity substances, including those used in nuclear technologies and electronics; and of ecological objects. In these areas, a great deal of practical experience has been accumulated.

8.2 Nuclear Materials

As is known, since the early 1940s, work on the use of atomic (more precisely, of course, nuclear) energy was almost simultaneously launched in the United States, the USSR, and Germany—at that time exclusively for military purposes. Almost

Y. A. Zolotov, *Russian Contributions to Analytical Chemistry*,
https://doi.org/10.1007/978-3-319-98791-0_8

immediately, the problem of creating materials for nuclear reactors and atomic bombs became of utmost importance. Constructional materials for nuclear reactors had to meet exceptional and previously never required standards. For example, zirconium was required not to contain any hafnium. Graphite with an exceptionally low ($<10^{-4}\%$) concentration of elements that strongly absorb neutrons—boron, cadmium, and some rare-earth elements—became necessary. Uranium had to satisfy stringent requirements, in terms of both its elemental and isotopic composition. Obtaining these, and other materials, to the required specifications would have been impossible without corresponding analytical procedures. Therefore, within the framework of "uranium projects" both in the USSR and the United States, special scientific units, responsible for providing analytical control and, accordingly, for scientific and technical solutions in this field, were created. The selection of specialists and the undertaking of research in this field was hampered by the secrecy of such "uranium projects." In the Soviet press, at one time it was even forbidden to mention the very word "uranium." To date, the scientific press has published small amounts of material on the chemical–analytical studies dealing with these projects.

In the USSR, A. P. Vinogradov (Fig. 8.1), future academician, was responsible for the "chemical–analytical" part of the atomic project. Many of the major analysts who took part in the atomic project, like Vinogradov, worked in the institutes of the USSR Academy of Sciences (P. N. Palei, D. I. Kurbatov, and many others). Some methodological work was carried out at scientific institutions in Moscow and other cities, e.g., at plant 817 (which would become the "Mayak" plant near Chelyabinsk) or at the Urals Electrochemical Combine. Groups of highly skilled analysts were brought together in the emerging nuclear industry, e.g., in the All-Union Institute of Inorganic Materials – the following future, or already accomplished, doctors of sciences, L. V. Lipis, V. K. Markov, S. V. Elinson, A. E. Klygin, I. V. Moiseev, A. V. Vinogradov, et al.; in the All-Union Institute of Chemical Technology – V. A. Pchelkin, Yu. K. Kvaratskheli, V. K. Lukyanov, and a number of others; and in the Physicotechnical Institute (Obninsk) – A. G. Karabash. Among the first

Fig. 8.1 Aleksandr Pavlovich Vinogradov (August 21, 1895–November 16, 1975) supervised the chemical–analytical direction of the Soviet atomic project. Photo provided by the Vernadskii Institute of Geochemistry and Analytical Chemistry

analysts involved in plutonium and other transuranium elements was Boris Vasil'evich Kurchatov, the brother of the scientific leader of the "uranium project," I. V. Kurchatov.

At the first stage of implementation of the "uranium project" in the Soviet Union, the search for uranium deposits began, this process required new analytical support. Thus, in 1945, production of gamma radiometers was set up; such devices were even installed on airplanes. In the development of this field, an important contribution was made by A. L. Yakubovich. Almost simultaneously, research was carried out into technology for obtaining and purifying uranium, separating and enriching its isotopes, and creating extremely clean construction materials for the construction of future nuclear reactors. The results of analyses were required at all stages of the project's implementation, with chemical–analytical studies being carried out in different research fields within the project at once.

Here is an excerpt from a top secret document [1], dated August 1946, sent to the government curator of the atomic project—L. P. Beriya. This is the report from the project organizers, B. L. Vannikov, I. V. Kurchatov, M. G. Pervuhin, I. I. Malyshev,

Fig. 8.2 The authors of the book on uranium determination have done a lot for analytical control in the nuclear industry

and I. K. Kikoin, about their work during a one-and-a-half-year period (1945–1946). Let us remind ourselves that the Soviet atomic bomb would be tested only three years from the date of this report.

> "Throughout the entire operation of the uranium-graphite reactor project, the development of sufficiently accurate methods of chemical analysis was paramount, very complex, and difficult to solve. The task of obtaining pure metal turned out to be very complicated, both because the technology of obtaining chemically pure uranium in general is very little developed, and because it is necessary to determine certain impurities in quantities … one two hundred thousandth [proportion] of percent with which our analytical laboratories and institutions in practice have not met. For the time elapsed in 1946 research institutes and laboratories involved in the development of the analytical method have found a way to determine the most dangerous impurities for the reactor and at present this complex research work is coming to an end … for all kinds of impurities. The chemical analysis of a number of batches of metallic uranium produced by the plant no. 12 showed that uranium can be obtained sufficiently clean, and thus there is every reason to expect that by the time the uranium-graphite reactor is built it will be provided with metal."

At that time various methods were used to determine the elemental composition of the materials of the atomic industry, including photometric (using organic reagents), polarography, and atomic emission analysis. In 1962, the book entitled *Analytical Chemistry of Uranium* was published, in which the methods of determining uranium developed at this time were systematized and generalized [2]. In it, in particular, it was noted: "Methods of determining uranium with uncolored reagents in recent years have not always met the increased sensitivity requirements. Therefore, there is a tendency to use organic colored reagents, as a rule, more sensitive. Of these, the reagent arsenazo 1 (or uranone), synthesized and proposed to determine uranium in 1941 by V. I. Kuznetsov was the most widely used" (Fig. 8.2).

Great experience in the analysis of raw materials for the nuclear industry, as well as compounds of uranium, beryllium, and other elements, was accumulated at the All-Union Institute of Chemical Technology. Here, as in other laboratories dealing with materials of atomic engineering, the role of chemical methods (including electrochemical and photometric methods) gradually decreased. The importance of methods of separation and concentration of elements (extraction, sorption, and ion exchange), which at first were used very widely, was reduced. However, the use of powerful physical methods was constantly expanding.

The development of methods for the analysis of uranium for neutron-absorbing and other impurities content greatly stimulated the creation and improvement of methods for determining the very low concentrations of any elements. Such work applied to atomic emission analysis was generalized in a book entitled *Spectral Analysis of Atomic Materials*, written by A. N. Zaidel, N. I. Kaliteevskii, L. V. Lipis, and M. P. Chaika (1960).

When studying the products of uranium fission, the problem arose of determining the large number of radioactive elements in their complex mixtures (radiochemical analysis). Radiochemical analysis was understood to be the means of identification, assessment of concentration, and sometimes determination of any physical characteristics of radionuclides in their mixtures. This kind of analysis was developed as early as the beginning of the 20th century, long before the uranium project. Later physicists encountered similar problems when deciphering the composition of products obtained by irradiating any elements or compounds in cyclotrons and synchrophasotrons. The solution to such problems was, and is, studied by radiochemists, as well as analysts. Within the framework of radiochemical analysis, a completely new problem arose—development of the analytical chemistry of transuranium elements, particularly plutonium, then neptunium, americium, etc. One idea which has arisen from time to time is to create a thorium reactor, something which has stimulated interest in protactinium.

Analysis of irradiated materials, media with high activity, is not an easy task not only from a scientific point of view, but also from the point of view of safety: radioactivity is radioactivity, of course. It is in this area that maximum automation is needed, including remote analysis. In this direction, much has been done in different countries around the world. Back in the late 1960s, progress achieved in this field was generalized in a book written by Shelemin [3].

In the 1940s–1950s, interest in ultramicroanalysis arose. This term was used to refer to a combination of procedures which involved operations with very small amounts of the sample—with solid samples weighing of the order of a few micrograms and with volumes around a microliter. Attention to ultramicroanalysis was stimulated by the tasks facing radiochemical analysis. Indeed, the masses of transuranium elements available in the first years of the atomic project realization! Studies on ultramicroanalysis were successfully developed in the United States (Kirk [4], Benedetti-Pikhler [5], and others), and in postwar years were continued in the USSR (I. P. Alimarin and M. N. Petrikova). At that time ultramicroanalysis used purely chemical methods. However, after a few years, ultramicroanalysis (in this purely chemical sense) lost its significance as a consequence of the emergence of more powerful physical methods.

An essential section of analytical research in this area was (and still is) isotopic analysis. For a long time it was only a method of scientific research. It was used by physicists, then by physicochemists, radiochemists, geologists (for determining the age of rocks), and archaeologists (using the carbon dating method). However, since solving the atomic problem at the beginning of the 1940s, it became a method of production control; this particularly was applied to the determination of uranium isotopes (^{235}U and ^{238}U). Another important area of isotopic analysis is control of the composition of "heavy water."

It became necessary to separate uranium isotopes (more precisely, to enrich uranium with its active isotope) on a large scale. In the USSR, a gas-diffusion method was used for this, then ultracentrifugation. The degree of isotope separation had to be carefully monitored. In the United States such work was conducted in a laboratory that later became known as the "Oak Ridge National Laboratory" (Tennessee State); in the USSR this work happened at the Urals Electrochemical Combine (Novoural'sk, Sverdlovsk Region). At this plant, a mass spectrometric laboratory for isotopic analysis of uranium hexafluoride has existed since 1948. This laboratory not only served the production, but also developed mass spectrometers. In addition to mass spectrometry for isotopic analysis, additional methods were used: isotopic decay, atomic-emission spectral methods, and subsequently, nuclear magnetic resonance. However, mass spectrometry is still the main method used for isotopic analysis. It is not by chance that at the beginning of the twenty-first century the Ministry of Atomic Energy of the Russian Federation ordered the Institute of Analytical Instrumentation of the Russian Academy of Sciences (RAS) to develop isotope mass spectrometers of a new generation.

At the Leningrad University, in 1948, the Laboratory of Spectral Analysis (A. N. Zaidel) was organized within the Department of Physics, its task was to develop and introduce sensitive methods which could be used for the elemental analysis of high-purity atomic materials; this work was reflected in the already mentioned monograph [3]. However, in the 1950 s the laboratory engaged in isotopic spectral analysis and achieved much in this direction (A. A. Petrov, A. G. Zhiglinskii, V. M. Nemets, et al.). Some books were written on this subject; one worth a mention was the monograph written by A. A. Petrov entitled *Spectral Isotopic Method of Material Research* (1974).

In the 1960s, civilian nuclear materials came to the forefront. This was, in particular, "nuclear fuel" for nuclear power plants. Since that time, state organizations responsible for nuclear energy have acted as customers and sponsors of research in the field of analytical chemistry (in Russia this falls to the state corporation Rosatom). A new field of research has opened up for analytical chemistry; new problems have emerged that require non-standard approaches, e.g., the task of controlling the composition of spent nuclear fuel. Interesting and complex problems also appeared in other research areas (detection of ultralow quantities of synthesized transcurium elements and detection of traces of nuclear explosions).

The analytic chemistry of transuranium elements was actively developed by the team at the radiochemical laboratory of the V. I. Vernadskii Institute of Geochemistry and Analytical Chemistry of the USSR Academy of Sciences (GEOKHI). This team was headed first by P. N. Palei, then—for many years—by B. F. Myasoedov (Fig. 8.3).

Fig. 8.3 Boris Fedorovich Myasoyedov and co-workers developed numerous methods for the analysis of nuclear materials. Photo provided by B. F. Myasoedov

***Boris Fedorovich Myasoedov** was born on September 2, 1930. He graduated from the D. I. Mendeleev Moscow Institute of Chemical Technology. He is an academician of the RAS, a professor, and worked as Deputy Director of the Vernadskii Institute and Deputy Chief Scientific Secretary of the RAS. He was an active member of the International Union of Pure and Applied Chemistry. Professor Myasoedov is a laureate of many awards, one of which was the international Heveshi Medal. Together with other laboratory staff, he developed numerous methods for isolating and determining protactinium, uranium, neptunium, plutonium, americium, and other elements. With the participation of others in his laboratory, he developed a precision coulometer which is used in the nuclear industry.*

8.3 Materials for Microelectronics and Other High-Purity Substances

It is difficult to talk about any major purely fundamental achievements of Russian analysts in this field. The results here are somewhat different: a successful (already mentioned) solution of difficult problems of analysis of uranium and other nuclear materials on impurities; development of methods and procedures for determining impurities in silicon, as well as in aluminum, gallium, arsenic, germanium, and other elements used in microelectronics; and a similar solution to problems in the manufacturing of optical fibers or heat-resistant alloys.

In the course of solving these problems new ideas were born, and non-standard techniques were created and disseminated. One of these widely used techniques was combining the preconcentration of trace elements, e.g., by evaporating extracts on a graphite powder, with the subsequent atomic emission determination of a large number of elements-impurities. These methods are called chemical–spectral methods.

The need to determine very low concentrations of a large number of impurities stimulated interest in methods which were capable, in principle, of solving such problems. Among such methods were, in addition to chemical–spectral methods, neutron activation analysis, spark source mass spectrometry, and many other methods which are widely used today. A large accumulated experience in the analysis of substances of high purity is to some extent imprinted within various publications, e.g. [6, 7].

A large set of carefully analyzed, high-purity substances collected in Nizhny Novgorod, formed an exhibition–collection at the G. G. Devyatykh Institute of Chemistry of High-Purity Substances of the RAS [8] (Fig. 8.4).

Another center, and a very large one at that, was the State Institute of Rare Metal Industry (Giredmet) in Moscow. Yu. A. Karpov headed the analytical department of this institute for many years, while V. V. Nedler, B. Ya. Kaplan, V. G. Goryushina, L. S. Vasilevskaya, and many others, actively worked at the institute in the fields of analysis of materials of electronics equipment and other pure substances.

Fig. 8.4 Book about the exhibition—collection of high-purity substances

A great contribution to the establishment and development of the analysis of high-purity materials in the USSR was made by I. G. Yudelevich, and the team which worked in the laboratory he created, in Novosibirsk. Combined analytical methods based on a combination of various preconcentration methods (liquid extraction, sorption, distillation, etc.) with highly sensitive instrumental methods—atomic emission, atomic absorption, mass spectrometry, and neutron activation analysis—were actively produced and developed in the laboratory. The laboratory continues its research and development today (A. I. Saprykin, I. R. Shelpakova, et al.).

Iosif Gershevich Yudelevich *(1920–1993) was a doctor of chemical sciences, a professor, and an Honored Worker of Science and Technology of the Russian Federation. He was born in Slonim, Poland, in 1920. After completing his studies at the high school, he studied at the physics and mathematics lyceum, and then in 1940 at the Physics and Mathematics Department of the Belarusian State University, where he studied till the war. He finished his studies in Alma-Ata, and his family—his parents and younger brother—died in Belarus in 1942. Yudelevich started his career at the Chimkent lead–zinc plant, and then worked as head of the laboratory of spectral analysis at the VNIItsvetmet Institute in Ust-Kamenogorsk. In 1963, he organized and headed a laboratory for monitoring the purity of semiconductor materials at the Institute of Inorganic Chemistry of the Siberian Branch of the USSR Academy of Sciences. Under the guidance of Yudelevich, more than 50 candidate of sciences theses were defended, and five of his trainees subsequently became doctors of science. Yudelevich headed the Siberian Branch of the Scientific Council of the USSR Academy of Sciences on Analytical Chemistry. Under his initiative in 1982, in Tyumen, the Analytics of Siberia and the Far East conferences were started, which became one of the largest conferences on analytical chemistry in Russia. He organized the Novosibirsk Analytical Seminar. Yudelevich also came up with the idea of organizing the Russian–German–Ukrainian Symposium (ARGUS), which also started in Novosibirsk.*

The V. I. Vernadskii Institute of Geochemistry and Analytical Chemistry of the USSR Academy of Sciences (RAS), where work has been done on the preconcentration of impurity elements and on the radioactivation analysis of electronic equipment materials, should also be named here; at the initial stage of such work, spark source mass spectrometry was used at the institute. Here, work was carried out under the supervision of A. P. Vinogradov and I. P. Alimarin, with the active participation of Yu. V. Yakovlev, M. S. Chupakhin, Yu. A. Zolotov, and many others (Fig. 8.5).

Ivan Pavlovich Alimarin *(September 11, 1903–December 17, 1989) was a general analyst, a member of the USSR Academy of Sciences, laureate of the USSR State Prize, and head of the Division of Analytical Chemistry at the M. V. Lomonosov Moscow State University (1953–1989) and the laboratory in the Vernadskii Institute. He headed the Scientific Council of the USSR Academy of Sciences on Analytical Chemistry and was the editor-in-chief of the Journal of Analytical Chemistry. He was highly respected abroad, was a member of the editorial boards of a number of journals, participated in the International Union of Pure and Applied Chemistry, and held a number of international awards.*

He has had work published on the analysis of mineral raw materials, substances of high purity, methods of separation and concentration, and radioanalytical methods. In many ways, he helped to set new promising directions and train analysts.

8.4 Rare Elements and Platinum-Group Metals

Sometimes analysts say that in the Soviet Union, in terms of the analytical chemistry of rare elements, more work has been published than in all the world's other countries combined. Perhaps this is true, especially in terms of rare-earth elements. Many procedures for determining almost all rare elements have been developed, using the full arsenal of analysis methods. Some of the procedures have been introduced into industrial, geological, and ecological laboratories, as well as other non-specific laboratories. There has been a lot literature devoted to the

Fig. 8.5 Ivan Pavlovich Alimarin headed the teams, who did a lot for the analysis of high-purity substances. Photo provided by I. P. Alimarin

analytical chemistry of rare elements, especially in the 1950s–1970s, e.g. [9, 10]. Certain volumes of the series *Analytical Chemistry of Elements*, published in 1970–1990 by the V. I. Vernadskii Institute of Geochemistry and Analytical Chemistry of the USSR Academy of Sciences (see, e.g. [9–13]), were devoted to many rare elements. Some analyst–researchers devoted almost all their scientific lives to the analytical chemistry of one element, as did L. V. Borisova, who studied rhenium. The Division of Analytical Chemistry of the M. V. Lomonosov Moscow University had a separate laboratory for analytical chemistry of rare elements, which was headed by professor A. I. Busev. Interest in these elements was triggered by their sharply increasing worldwide use linked to various areas of the economy or security.

On the contrary, interest in platinum metals originated in Russia quite a long time ago. By the eighteenth to nineteenth centuries much had been done in terms of their isolation, detection, and quantitative determination (P. G. Sobolevskii, C. C. Claus, et al.) (Fig. 8.6).

In 1840–1850, C. C. Claus, who discovered ruthenium, developed some methods for the isolation and detection of platinum metals [14]. He used separate compounds of these elements as analytical reagents. A book was been devoted to this work [15]. In the twentieth century, a large number of studies considered the problem of analytical control of the extraction technologies used to obtain platinum metals from the raw materials of the Norilsk Mining and Metallurgical Combine, with subsequent refinement of metals at the Krasnoyarsk Non-Ferrous Metals Plant and additional processing at the Sverdlovsk (Yekaterinburg) Non-Ferrous Metal Processing Plant. A significant part of the research work, implemented in the control and analytical laboratories of these enterprises, was performed at the N. S. Kurnakov Institute of General and Inorganic Chemistry of the USSR Academy of Sciences in 1940–1960, and later at the A. V. Nikolaev Institute of Inorganic Chemistry of the Siberian Branch of the USSR Academy of Sciences (Novosibirsk), the M. V. Lomonosov Moscow Institute of Fine Chemical Technology, and many institutes of industry—Giredmet, Gipronickel, TsNIGRI, etc.

Fig. 8.6 Carl Carlovich Claus (January 22, 1796–March 24, 1864) discovered ruthenium and developed several methods for determining platinum metals. Picture previously published

An important role in the development of practically useful procedures for determining platinum metals was played by the already mentioned State Institute of Rare-Metal Industry (Giredmet), in which the Department of Analytical Chemistry was led for a long time by Yu. A. Karpov (Fig. 8.7).

Yurii Alexandrovich Karpov *is born on March 1, 1937. He is an academician of the RAS and a professor. He graduated from the Moscow Institute of Steel and Alloys, worked at the Institute of Metallurgy of the USSR Academy of Sciences, mainly in Giredmet. Currently he is the Principle Research of the N. S. Kurnakov Institute of General and Inorganic Chemistry. He was deputy chairman of the Scientific Council of the RAS on Analytical Chemistry, editor-in-chief of the journal Plant Laboratory. Diagnosis of Materials. He is the president of the "Analytics" Association.*

He carried out work on the determination of gases in metals, analysis of high-purity substances, analytical chemistry of rare and platinum metals, and analytical metrology and standardization.

All this accumulated experience was reflected in a number of books [16, 17] (Fig. 8.8). In addition, this experience was shared at regularly held conferences known as the *Chernyaev Conferences on Chemistry, Technology, and the Analysis of Platinum Metals* (I. I. Chernyaev was the director of the N. S. Kurnakov Institute of General and Inorganic Chemistry, an expert in complex compounds of platinum metals).

8.5 Environmental Objects

This group of objects began to occupy a leading position in many laboratories since 1970, with interest peaking in the 1990s. A lot of useful procedures, specialized techniques, standard reference samples, etc., have been created. Many university departments, institutes of the Academy of Sciences, and control analytical laboratories are involved in such work. A significant amount of experience has accumulated, e.g., in analysis of water by ROSA, the Analytical Center for Water Quality Control in Moscow.

Fig. 8.7 Yurii Aleksandrovich Karpov led work on the analysis of high-purity substances and rare metals at the Giredmet Institute. Photo provided by Yu. A. Karpov

Fig. 8.8 Several books on the analysis of specific objects and on the analytical chemistry of platinum metals

Attempts were made to formulate a common methodology for monitoring environmental objects from the point of view of the chemical analysis. One of the schemes [18] is set out below.

The main task of analysis and control of environmental objects is, of course, the detection, identification, and quantification of harmful substances of anthropogenic origin. However, a problem arises also in terms of the control of natural components, such as, e.g., carbon dioxide and ozone in the atmosphere or dissolved oxygen in water.

The control of environmental objects, in terms of determining their chemical composition, has encountered great difficulties, which are in part associated with the features of the environmental objects themselves, e.g., with the diversity of their natural composition and lability, variability in time. Other difficulties are associated with the magnitude of the task, the sheer amount of work required. In fact, the number of substances which need to be controlled is growing. For example, it is known that for natural water, Russia has established maximum permissible concentrations (MPCs) for approximately 1500 substances. The number of required analyses, even for the most common pollutants, is increasing. As time progresses new types of objects, which are considered desirable to control, appear. All this is remarkable since it reflects society's concerns about the cleanliness of habitats, improvements to quality of life, etc. However, for this to happen the responsibilities analysts and controlling organizations (worldwide), should rapidly increase.

The matter is complicated by the fact that the common methodology of analysis and control of environmental objects, which is often used today, does not make it possible to solve more complicated tasks, e.g., reliably determining all 1500 substances in natural water. Let us assume that reliable methods are eventually developed. We still have the problem that to date no monitoring laboratory can determine all these substances simultaneously—not least because of the enormous amount of effort required. Of course, we understand that in reality there is no need to determine all these substances simultaneously, but this is another matter. What do we have in reality? We have good domestic laboratories which can determine 100–150 indicators, conventional laboratories which can determine 20–30 indicators, and other laboratories which can determine fewer.

The way out, apparently, is to change the methodology itself—the general approaches to analysis and control. The concept of monitoring environmental objects in terms of determining their chemical composition may include several components; from these we shall consider just six. It should be emphasized that these elements are closely interrelated.

The provision of a large-scale, two-step analysis (or even a three-step analysis) with screening of samples at the first stage can be considered the first such component. Screening greatly facilitates and simplifies control. During this first stage, samples which yield a negative result for a semi-quantitative, simplified assessment of the presence or absence of an initial substance are rejected. A negative result, i.e., data on the absence of the initial substance or, more precisely, on its possible presence below a prescribed concentration limit, is considered to be final, and these samples are no longer returned to. Samples which give a positive result go to the second stage of the analysis, in which more precise methods are used. This first stage is inexpensive and convenient, because the task here is to provide large-scale and express evaluation.

The second important element of the methodology can be considered as much broader use of generalized, integrated indicators. Here, we move away from laborious component analysis. Generalized indicators are especially valuable to the first stage of control. It is important to expand the search for new generalized indicators for inclusion in the system of multistage analysis. In the future, apparently, it will be possible to exclude known techniques from use, or perhaps improve them (chemical oxygen consumption, total organic carbon, tests for the sum of heavy metals, etc.). Special attention should be paid to the creation and improvement of biotests, especially when applied to water. They should be fast to use, cheap, and based on the idea of being easily accessible whilst providing long-term storage and transportation of organisms. Such biotests should be operable in the field (so-called simple tests). Existing rules also require that they be certified. Such biotests would need to be used at the very first stage of control.

Another way to get away from component-wise analysis—this time almost entirely—is the use of pattern recognition. Recognition of the general image of the object of research, primarily with the help of electronic tongue and electronic nose technology, is rarely used. However, with commercial development this technique will be more widely used in the field, e.g., an electronic tongue will certainly be able to indicate a change in the composition of water.

An obvious and successful element of the concept under discussion is the movement of chemical analysis from a stationary laboratory to the location of the analyzed object, i.e., in the field. The advantages of non-laboratory analysis are obvious. Time and money are saved in terms of a sample's preservation and transportation. You can analyze samples practically, in situ, in real time. For analysis in the field, simpler and cheaper means are often used than in a laboratory, and, consequently, the requirements for realizers are reduced. More importantly, there are objects that are difficult to analyze in the laboratory. This applies to labile specimens and to emergency situations, including those of natural origin. Of course, the most complex analyses of the most complex objects in the field can not yet be carried out.

In order to providing large-scale analysis, especially in the field, it is important to create and widely use mobile means of analysis. In principle, the arsenal of mobile means includes mobile laboratories in cars, boats, railway carriages, helicopters, and airplanes. Further it includes portable devices, including those which are pocket sized; chemical and biochemical test kits, test systems, as well as chemical sensor systems. In this line, mobile laboratories are, of course, a palliative tool, intermediate between stationary laboratories and systems of portable devices.

An essential condition for large-scale analysis is, of course, automation. One variant is the creation of continuous mode stations and posts which operate automatically to monitor the composition of air and water. However, the most toxic organic compounds can not be determined in this way. Another variant is the automation of large-scale laboratory analysis, primarily of water. Much has been done, in particular, to develop methods based on flow-injection analysis and its analogs. This is undoubtedly a promising way forward. Let us not forget that many modern devices are largely automated by themselves.

A large volume of work has been carried out to control persistent organic substances [19], e.g., dioxins, at the Severtsov Institute of Ecology and Evolution, RAS (this work began as a consequence of the Vietnam War and the use of Agent Orange).

8.6 Other Analysis Objects and Analytes

Much attention has been paid and is given to the analysis of mineral raw materials, geological, and geochemical objects. In this area, a lot of experience has been accumulated, with many leading analysts of the twentieth century starting in such laboratories, especially at the All-Union Institute of Mineral Raw Materials (VIMS) in Moscow. Many methods of analysis, new for their time, were "run-in" in such institutes (polarography, nuclear physics methods, atomic absorption spectrometry, etc.). The same is true for the analysis of metals and alloys, where the main analytical problems are currently solved by spectrometric methods.

As in many other countries, new methods for the analysis of food and pharmaceuticals are being developed. In recent years, leading laboratories have been creating new methods for analyzing biomedical objects, including for the purposes of diagnosing diseases. In 2015, a major conference entitled *Chemical Analysis and Medicine* was held in Moscow.

8.7 Ensuring the Quality of Chemical Analysis

In the USSR, the state service of standard reference materials (SRM), which organizes the development and production of standard samples of substances and materials, was established. The scientific and methodological center of the service was in Yekaterinburg, which was then called Sverdlovsk. An important role in organizing this standardization was played by A. B. Shaevich (1924–2017), the author of several books, including one entitled *Analytical Chemistry as a System* [20] (Fig. 8.9).

The development and production of SRM of ferrous metals, alloys and, in general, materials of ferrous metallurgy (slags, fluxes, etc.) were well organized. This required a lot of involvement from one of the branch institutes of ferrous metallurgy, located in Sverdlovsk–Yekaterinburg. Numerous research institutes of non-ferrous metallurgy, e.g., the institute located in Mtsensk, created standard samples of copper alloys and other products of non-ferrous metallurgy. Several institutes of the geological service, as well as three or four institutes of the Academy of Sciences, created, and continue to create, SRM for corresponding natural objects,

Fig. 8.9 Aron Borisovich Shaevich (October 24, 1924–2017) contributed, in many ways, to the creation of a system of standard samples of composition. Photo provided by A. B. Shaevich

e.g., the A. P. Vinogradov Institute of Geochemistry has worked successfully in this field. The D. I. Mendeleev All-Russian Institute of Metrology (St. Petersburg) is involved in standard gas mixtures. The "Ekoanalitika" association in Moscow produces SRM of a number of environmental objects.

Many analysts are engaged in the traditional metrology of chemical analysis—not so much in the scientific sense, but in terms of generalization, adaptation, popularization, and training. Work on chemometrics in terms of processing multidimensional data is very impressive (A. L. Pomerantsev, O. E. Rodionova, B. M. Mar'yanov, V. I. Vershinin et al.).

The association of analytical centers "Analitika" has been engaged for many years with the accreditation of laboratories. General requirements for the certification of laboratories are based on the recommendations of the International Organization for Standardization (ISO) and the Russian government organization Rosstandart.

Readers are directed to the book written by V. I. Dvorkin entitled *Metrology and Quality Assurance of Chemical Analysis* [21].

References

1. Ryabev, L.D. (ed.): *Atomic Project of the USSR: Documents and Materials, Atomic Bomb 1945–1954*. Nauka-Fizmatlit, Moscow, 552 pp. (2000) (in Russian)
2. Markov, V.K., Vernyi, E.A., Vinogradov, A.V. et al.: *Uranium: Methods of Its Determination*, 2nd edn. Atomizdat, Moscow, 502 pp. (1964) (in Russian)
3. Shelemin, B.V.: *Automatic Analyzers of Radiochemical Media*, 2nd edn. Atomizdat, Moscow, 294 pp. (1965) (in Russian)
4. Kirk, P.: *Quantitative Ultramicroanalysis*, translated from English, edited by I.P. Alimarin. Izd-vo inostr. lit., Moscow, 376 pp. (1952) (in Russian)
5. Benedetti-Pikhler, A.: *The Technique of Inorganic Microanalysis*, translated from English, edited by I.P. Alimarin. Izd-vo inostr. lit., Moscow, 296 pp. (1952) (in Russian)
6. Alimarin, I.P. (ed.): *Methods for Analysis of High-purity Substances*. Nauka, Moscow, 532 pp. (1965) (in Russian)
7. Alimarin, I.P. (ed.): *Methods for Analysis of High-purity Substances* (Problems of Analytical Chemistry No. 7). Nauka, Moscow, 311 pp. (1987) (in Russian)
8. Devyatykh, G.G., Karpov, Y.A., Osipova, L.I.: *The Exhibition–Collection of Substances of Special Purity*. Nauka, Moscow, 236 pp. (2003) (in Russian)
9. Ryabchikov, D.I., Ryabukhin, V.A.: *Analytical Chemistry of Rare-Earth Elements and Yttrium* (Analytical Chemistry of Elements Series). Nauka, Moscow, 328 pp. (1966) (in Russian)
10. Busev, A.I., Tiptsova, V.G., Ivanov, V.M.: *Manual on Analytical Chemistry of Rare Elements*. Khimiya, Moscow, 431 pp. (1978) (in Russian)
11. Dymov, A.M., Savostin, A.P.: *Analytical Chemistry of Gallium* (Analytical Chemistry of Elements Series). Nauka, Moscow, 256 pp. (1968) (in Russian)
12. Plyushchev, V.E. and Stepin, B.D., *Analytical Chemistry of Rubidium and Cesium* (Analytical Chemistry of Elements Series). Nauka, Moscow, 224 pp. (1975) (in Russian)
13. Borisova, L.V., Ermakov, A.N.: *Analytical Chemistry of Rhenium* (Analytical Chemistry of Elements Series). Nauka, Moscow, 310 pp. (1974) (in Russian)

14. Claus, C.: *Beiträge zur Chemie der Platin-metalle* (Festschrift zur jubelfeier des fünfzigjahrinen Bestehens der Univesität Kazan, 1854 [written to commemorate the 50th anniversary of the foundation of Kazan University]). University Press, Kazan
15. Claus, C.C.: *Selected Works on the Chemistry of Platinum Metals*. Nauka, Moscow, 304 pp. (1954) (in Russian)
16. Ginzburg, S.I., Ezerskaya, N.A., Prokof'eva, I.V. et al.: *Analytical Chemistry of Platinum Metals*. Nauka, Moscow, 614 pp. (1972) (in Russian)
17. Zolotov, Y.A., Varshal, G.M., Ivanov, V.M. (eds.): *Analytical Chemistry of Platinum Group Metals*. URSS, Moscow, 592 pp. (2003) (in Russian)
18. Zolotov, Y.A.: Zh. Anal. Khim. **65**, 227 (2010)
19. Maistrenko, V.N., Klyuev, N.A., *Ecologo-Analytical Monitoring of Persistent Organic Pollutants*. Binom. Lab. Znanij, Moscow, 323 pp. (2004) (in Russian)
20. Shaevich, A.B.: *Analytical Service as a System*. Khimiya, Moscow, 264 pp. (1981) (in Russian)
21. Dvorkin, V.I.: *Metrology and Quality Assurance of Chemical Analysis*. MITKhT, Moscow, 424 pp. (2014) (in Russian)

Chapter 9
Textbooks, Journals, History, and Methodology of Analytical Chemistry. The Means of Promotion of This Science

Abstract As early as the nineteenth century, textbooks on analytical chemistry (N. A. Menshutkin and F. F. Beilstein) were written in Russia, which have been published many times in many different languages. Modern textbooks have been produced at Moscow and St. Petersburg Universities as well as Moscow Technological University (editors, respectively, Yu. A. Zolotov, L. N. Moskvin, and A. A. Ishchenko). Four journals on analytical chemistry are currently published, including the *Journal of Analytical Chemistry* published simultaneously in English and Russian. Many publications deal with the history and philosophy of analytical chemistry, including the book by Yu. A. Zolotov and V. I. Vershinin entitled *History and Methodology of Analytical Chemistry*.

9.1 General Remarks

The success of any natural science is largely determined by the availability of necessary instrumentation, reagents, etc., and the rational organization of work (or, in other words, having suitable financial, material, and administrative conditions). However, the main driver for the successful development of science is, of course, people—erudite, creative, motivated, responsible, and hardworking. The system of training scientific personnel should ideally be aimed at cultivating and selecting such researchers.

Professional training of analysts includes many components and requires the use of various forms and means. One tools available is, of course, textbooks, as well as various auxiliary literature—handbooks, practical guides, etc.

Students and teachers, as well as actively working researchers should also be familiar with the history of their chosen science, understand the logic of the development of ideas, methods, and procedures of work, and a knowledge of the founders and the discoverers within their field. V. I. Vernadskii wrote that studying the history of science is a tool for creating a new one.

For the development of a specific field of science, it is essential that is has support from society and an understanding by the state, and broad mass of people,

Y. A. Zolotov, *Russian Contributions to Analytical Chemistry*,
https://doi.org/10.1007/978-3-319-98791-0_9

that it is important and useful. This represents the cornerstone which supports science, enabling a flow of people, and financial support, into it. Hence the desirability of propaganda and popularization of science. This also applies, of course, to analytical chemistry.

9.2 Textbooks and Handbooks. Serial Publications

The first manuals on analytical chemistry appeared at the beginning of the nineteenth century. Thus, in 1801, V. M. Severgin (1765–1826) published the book *Assay Art, or Guide to the Test of Metal Ores*. In 1854, N. A. Ivanov (1816–1883), who worked at the St. Petersburg Mining Institute, published a three-volume manual on chemical analysis, entitled *Initial Basis of Analytical Chemistry*, which was awarded the Demidov Prize. In 1867, F. F. Beilstein, who worked at the St. Petersburg Technological Institute, wrote, together with L. Yu. Yavein, a book entitled *A Guide to Qualitative Chemical Analysis*. At the same time, the book was published in German in Leipzig, in 1868 was translated into Dutch, in 1873—into English, and in 1876—French. In Russia, the "Guide" saw nine editions (the last being published in 1909) and for several decades served as an important textbook in analytical chemistry. Fedor Fedorovich Beilstein (1838–1906), known for his handbook on organic chemistry, was a member of the St. Petersburg Academy of Sciences. N. A. Menshutkin's textbook entitled *Analytical Chemistry*, published in its first edition in 1871 and later reprinted in Russia a further 15 time (until 1932!), it was also translated into the main European languages. Nikolai Aleksandrovich Menshutkin (1842–1907) was in charge of the Division of Analytical Chemistry at St. Petersburg University, and at the end of his life he also worked at the St. Petersburg Polytechnic Institute, of which he was one of the founders (Fig. 9.1).

Fig. 9.1 Nikolai Aleksandrovich Menshutkin (October 12, 1842–January 23, 1907) was in charge of the Division of Analytical Chemistry of St. Petersburg University and wrote a textbook entitled *Analytical Chemistry* which was published in Russian 16 times and translated into all the main European languages. Photo previously published

In the middle of the twentieth century, V. N. Alekseev's textbooks, translated into a number of languages, were popular; they were particularly widely used in developing countries and countries of Eastern Europe. These textbooks (*Qualitative Analysis*, *The Course of Qualitative Chemical Semimicroanalysis*, and *Quantitative Analysis*) were distinguished by their successful construction, clarity of presentation, and clarity of formulations, but dealt with only chemical methods of analysis.

Among the modern textbooks one should consider those prepared at Moscow University [1], St. Petersburg University [2] (Fig. 9.2), and the Moscow University of Fine Chemical Technology [3].

In terms of handbooks, the most well-known was Yu.Yu. Lurie's book entitled *Handbook of Analytical Chemistry*.

Among the serial publications are: *Analytical Chemistry of Elements*, *Analytical Chemistry Methods*, *Analytical Reagents*, and *Problems of Analytical Chemistry*. Many editions from these series have references in this book.

Fig. 9.2 A textbook on analytical chemistry, edited by L. N. Moskvin

9.3 Journals

The first specialized scientific journal, in which analytical chemistry took an important place, was the journal entitled *Plant Laboratory*, created in 1932. It is now published under the name *Plant Laboratory. Diagnosis of Materials (Zavodskaya Laboratoriya)*. In addition to the problems of chemical analysis, the journal considers the testing of physical and mechanical properties and mathematical methods.

The main scientific journal on analytical chemistry in Russia is the *Journal of Analytical Chemistry* (*Zhurnal Analitichescoi Khimii*), which has been published by the Academy of Sciences since 1946. The journal simultaneously appears in Russian- and English-language versions (the latter is distributed by Springer Publishing Company). Original articles, reviews, so-called general articles, information materials, and information on the history or teaching of analytical chemistry are published within this journal.

In Yekaterinburg there is a journal published under the name *Analytics and Control*, in Voronezh *Sorption and Chromatographic Processes*. Advertising and information magazine *Analytics* is published in Moscow (Fig. 9.3).

Fig. 9.3 The journal *Analytica*

Articles on analytical chemistry are published in many other journals—*Reports of the Academy of Sciences* and *Izvestiya RAN. Chemical series, Radiochemistry*, and *Izvestiya Vysshikh Uchebnykh Zavedenii. Chemistry and Chemical Technology*, *Chemistry for Sustainable Development*, etc.

9.4 Work on the History and Methodology of Analytical Chemistry

A few people were and are professionally engaged in the history of analytical chemistry in Russia, including Batalin [4], Tsyurupa [5], Ushakova [6], Strel'nikova [7, 8], and Zolotov and Apenova [9], all of whom have devoted their research mostly to domestic history. It is especially necessary to note the great work of E. M. Senchenkova, on the history of the creation of chromatography by Tswett [10–13]. L. B. Pavlova and A. N. Shamin produced a list of publications devoted to the history of analytical chemistry [14]. Yu. A. Zolotov and V. I. Vershinin had a chronology published of the most important discoveries in the field of analytical chemistry [15].

The general methodology ("philosophy") of analytical chemistry was one of the favorite topics of several Russian analysts—Yu. A. Klyachko, A. B. Shaevich, V. I. Vershinin, and Yu. A. Zolotov. In their articles and books, they considered names of the science on determination of chemical compositions, definitions, developmental stimuli, the relationships between fundamental and applied aspects of analytical chemistry, analytical chemistry's place in the system of sciences, the connection between analytical chemistry and the other sciences, and the hybridization of methods of analysis. Yu. A. Zolotov and V. I. Vershinin, in 2007, published the book published entitled *History and Methodology of Analytical Chemistry* [15].

One can consider this direction in more detail, in essence.

Periodically flaring discussions about the most common problems of analytical chemistry related to its self-knowledge have made it possible, although slowly, to move toward a coherent understanding of the basic concepts; to the formulation of more and more precise definitions, to finding terms acceptable for most; to a solution, although sometimes palliative, of a number of controversial issues, e.g., whether analytical chemistry is an applied science or whether it is proper to attribute it to the fundamental sciences.

The range of methodological issues includes, at a minimum, the following: the definition of analytical chemistry (including the concepts of purpose, subject, method, and relationship); incentives for its development; its position in the system of sciences (including its historical position, and most importantly its position in terms of interrelations and boundaries with other sciences); the routes to creating methods of analysis (an issue closely related to the previous one); the relationship between analytical chemistry and analytical service, and whether it is fundamental science or an applied science; and finally, the problem of the very name which should be given to a science which determines chemical compositions.

The common view is that these issues should not concern researchers, who, of course, have too many other—concrete—things to do; let science theorists do this, if it interests them; we have no time for such discussions. Meanwhile, such a position can hardly be considered correct (by the way, it reflects not so much the incredible employment of serious people who solve critical analytical problems, as inability to rise above everyday tasks and reach the level of the "philosophy" of science). The point is that the position of analytical chemistry in the system of knowledge, and even in society, depends to a certain extent on the coordinated, "correct" understanding of a number of these, and other unspecified, common problems; to some extent the very fate of our science and the scope of our efforts.

Let us consider one example. In the structure of one institute or one enterprise, under one roof, one can see the laboratory of analytical chemistry and the laboratory of spectral analysis. This means that the leaders of the institute, or the enterprise, do not understand that modern analytical chemistry includes spectral analysis, like many other directions and methods for studying chemical compositions. Now, analysts—chemists, physicists, and engineers—working in these laboratories, do not explain this to their leaders. They do not explain to their superiors that both words in the term "analytical chemistry" have meaning. It is not necessary to prove that today analytical chemistry uses more physical methods than chemical ones, e.g., optical spectral, X-ray, mass spectrometric, and nuclear–physical (radioactivation), as well as methods based on thermal conductivity (catharometers) and other concepts. The chiefs, apparently, believe that analytical chemistry purely incorporates chemical methods of analysis.

Does the community of analysts suffer because of this misunderstanding? And how do they suffer! Perhaps from an artificial disunity? Meanwhile, there is a lot which should unite specialists using different methods: rational distribution of the objects of analysis "by methods", solving of issues about the interchangeability and complementarity of methods, comparing results, a uniform solution of many metrological problems, etc.

And what is the opinion of certain science figures that analytical chemistry is called upon to serve, including "their" areas of knowledge, that this is not an independent scientific discipline. They do not know the difference between analytical chemistry as a science and analytical service as a system for providing analyses, including related and non-adjacent sciences. It's just as if some philistine under the chemistry would understand only the production of fertilizers for his summer residence and good detergents for his beloved wife. Such opinions can not simply be ignored, because their supporters speak at meetings of academic councils, vote, some of them occupy important positions, at which organizational decisions are made. Is it not from this that often the closing of the divisions of analytical chemistry in universities or the appointment to manage them by not analysts? In analytics, as in literature or medicine, everyone understands….

These and similar issues were considered at various seminars and conferences, including two Russian conferences on the history and methodology of analytical chemistry.

9.5 Promotion and Popularization of Analytical Chemistry

The popularization of, and generation of propaganda for, analytical chemistry is given little attention (in the United States it has gained more attention than in other countries). In Russia, such work is conducted, as elsewhere, on a very small scale, something which should be pitied.

It is possible to note separate articles which have appeared in newspapers (*Poisk*, etc.) and large-scale journals (*Chemistry and Life*, *Science and Life* (in Russian)) as well as appearing in one popular book by Yu. A. Zolotov entitled *Chemical Analysis for All, All, All* (in Russian) (2012) [16] (Fig. 9.4).

Fig. 9.4 One popular scientific book on chemical analysis

References

1. Zolotov, Y.A. (ed.): Fundamentals of Analytical Chemistry, in 2 parts, 6th ed., vol. 1, 400 pp.; vol. 2, 416 pp. IC Academiya, Moscow (2014) (in Russian)
2. Moskvin, L.N. (ed.): Analytical Chemistry, in 3 parts, vol. 1, 2008, 576 pp, vol. 2, 2008, 304 pp, vol. 3, 2010, 368 pp. IC Academiya, Moscow (in Russian)
3. Ishchenko, A.A.: Analytical Chemistry and Physicochemical Methods of Analysis, in 2 parts, vol. 1, 352 pp., vol. 2, 416 pp. IC Academiya, Moscow (2010) (in Russian)
4. Batalin, A.Kh.: Analytical Chemistry and Ways of its Development (History of the Origin and Development of Basic Methods and Directions of Analytical Chemistry). Tr. Orenburg. Sel'khoz. In-ta, 12, 368 pp. Orenburg (1961) (in Russian)
5. Tsyurupa, M.G.: A Short Essay on the Development of Analytical Chemistry, Part 1, Development of Analytical Chemistry Before the Beginning of the 20th century, 42 pp. Moscow State University, Moscow (1976) (in Russian)
6. Ushakova, N.N., Figurovskii, N.A.: Vasilii Mikhailovich Severgin, 160 pp. Nauka, Moscow (1981) (in Russian)
7. Strel'nikova, E.B.: Vopr. Ist. Estesvozn. i Tekhn., **4**, 48 (1986)
8. Savvin, S.B., Strel'nikova, E.B.: Zh. Anal. Khim. **38**, 727 (1983)
9. Zolotov, Yu.A., Apenova, S.E., Kara-Murza, S.G.: The Birth and Development of Methods of Chemical Analysis, 31 pp. Izd-vo Znanie, Moscow (1991) (in Russian)
10. Senchenkova, E.M.: Mikhail Semenovich Tswett, 307 pp. Nauka, Moscow (1973) (in Russian)
11. Senchenkova, E.M.: From the Collector. In: Tswett M.S. Selected Works, Ed. by Yu.A. Zolotov. The compiler, the author of sketches and comments E.M. Senchenkova, Moscow: Nauka, 2013, 679 pp. (in Russian)
12. Senchenkova, E.M.: The Birth of the Idea and the Method of Adsorption Chromatography, 448 pp. Nauka, Moscow (1991) (in Russia)
13. Senchenkova, E.M.: M.S. Tswett - The Creator of Chromatography, 440 pp. Janus-K, Moscow (1997) (in Russian)
14. Pavlova, L.B., Shamin, A.N.: Zh. Anal. Khim. **47**, 230 (1992)
15. Zolotov, Yu.A., Vershinin, V.I.: History and Methodology of Analytical Chemistry, 464 pp. IC Academiya, Moscow (2007) (in Russian)
16. Zolotov, Yu.A.: Chemical Analysis for All, All, All, 232 pp. GEOS, Moscow (2012) (in Russian)

Chapter 10
Organizational Input

Abstract Many Russian analysts have worked at the International Union of Pure and Applied Chemistry (IUPAC), the European Association of Chemical and Molecular Sciences (EuCheMS), the European Commonwealth of Metrology in Analytical Chemistry (EURACHEM), and other international organizations. Well-known Russian scientists sit on the editorial or advisory boards of a number of leading analytical journals. Several large international conferences have been held in Russia. Coordination of research by Russian analysts is carried out by the Scientific Council on Analytical Chemistry, which is a member of EuCheMS.

10.1 General Remarks

Russian specialists in analytical chemistry are known in scientific, organizational, and publishing communities.

Russia plays an important role in international organizations, editorial boards of international journals, and the organization of international conferences. In addition, Russian analysts also lecture in other countries. Several Russian analysts, having moved to other countries, have gone on to undertake significant work in those countries.

10.2 Work in International Organizations

Russian analysts work in many international organizations which are connected to chemistry or, more specifically, to analytical chemistry. Among these organizations are the International Union of Pure and Applied Chemistry (IUPAC), the European Association of Chemical and Molecular Sciences (EuCheMS), the European Commonwealth of Metrology in Analytical Chemistry (EURACHEM), the Cooperation on International Traceability in Analytical Chemistry (CITAC) (Fig. 10.1).

Y. A. Zolotov, *Russian Contributions to Analytical Chemistry*,
https://doi.org/10.1007/978-3-319-98791-0_10

Fig. 10.1 Activists of the Division of Analytical Chemistry of the Federation of European Chemical Societies (now EuCheMS). From left to right: Ya. Labuda, J. Horvai, Yu. A. Zolotov, S. Grob, V. N. Zaitsev, and M. Karayannis (1990s). Picture provided by V. N. Zaizev (Ukraine)

In the Analytical Chemistry Division of IUPAC, Russian analysts have been involved since the 1960s; the first of whom, apparently, was I. P. Alimarin, who was a member of the committee of the Analytical Chemistry Division and also a member of at least one commission, i.e., the commission on the nomenclature of analytical chemistry. Yu. A. Zolotov, was twice a member of the committee of the Analytical Chemistry Division and worked there from 1971 to 1987; he was also a member of several commissions. S. B. Savvin, Yu. G. Vlasov, E. A. Terent'eva, E. E. Gelman, B. Ya. Spivakov, V. P. Kolotov, T. A. Maryutina, P. S. Fedotov, and many others were among the members of the Analytical Chemistry Division. P.S. Fedotov was elected chairman of the "Chemistry and Ecology" Division of IUPAC.

For a long time Yu. A. Zolotov worked in the Analytical Chemistry Division of the Federation of European Chemical Societies (FECS), at that time named EuCheMS; he represented the Russian Chemical Society and received a tribute for this work in 2012 (Fig 10.1). Then he was succeeded by S. N. Shtykov. After joining EuCheMS representing the Scientific Council of the Russian Academy of Sciences in Analytical Chemistry, B. Ya. Spivakov became an active member of the board. S. N. Shtykov currently heads the working group on nanotechnology at the Analytical Chemistry Division [1].

At *Euoranalysis* conferences, regularly organized by the Division of Analytical Chemistry of EuCheMS, scientific committees have included Yu. A. Zolotov, B. Ya. Spivakov, S. N. Shtykov, O. A. Shpigun, and others.

The Association of Analytical Centers “Analytics” (discussed below) is a member of the EURACHEM. In this community Russia is represented by V. B. Baranovskaya and M. Yu. Medvedevskikh. “Analytics” also interacts with CITAC, whose Russian representative is Yu. A. Karpov. “Analytics” is part of the Asia–Pacific Cooperation for Laboratory Accreditation (APLAC) and has contacts with the International Laboratory Accreditation Cooperation (ILAC), see [2].

10.3 Work in International Journals

In 1970–1990, Russian analysts worked in boards of most leading journals on analytical chemistry, except, perhaps, *Analytical Chemistry*. These journals include: *Analytica Chimica Acta, Talanta, Analytical Letters, Fresenius Zeitschrift főr analytische Chemie, Microchimica Acta, The Analyst, Critical Reviews in Analytical Chemistry*.

In addition to these wide-ranging publications, Russian analysts sat on the editorial boards of many specialized journals: *Journal of Chromatography, Chromatographia, Spectrochimica Acta, Journal of Radiochemical and Nuclear Chemistry, Electroanalysis, International Journal of Environmental Analytical Chemistry, and other*.

Examples of Russian scientists on the editorial boards of journals include: E. I. Morosanova, a member of the editorial board of the *Talanta*; V. G. Berezkin, *Chromatographia*; Yu. G. Vlasov, *Sensor and Actuators* and other journals; Yu. A. Zolotov, *Analytica Chimica Acta, International Journal of Environmental Analytical Chemistry, Talanta*, and other journals; A. I. Busev and A. M. Egorov, *Analytical Letters*.

10.4 Organization of International Conferences

Here are some major international conferences organized in Russia:

1. *Conference on the Application of Radioactive Isotopes in Analytical Chemistry* (Moscow, 1957).
2. *Analytical Section of the International Congress on Pure and Applied Chemistry* (Moscow, 1965).
3. *International Conference on Solvent Extraction* (Moscow, 1988).
4. *International Symposium on Kinetics in Analytical Chemistry* (Moscow, 1995).
5. *International Congresses on Analytical Science* (Moscow, 1997, 2006) (Fig. 10.2).
6. *Conference on the Centenary of Chromatography* (2003).
7. *USSR–Japan Symposia on Analytical Chemistry* (1984, 1988, 1992, and 1996).
8. *Russian–Ukrainian–German Symposia on Analytical Chemistry* (1986–2007).

Fig. 10.2 In the Presidium of the International Congress on Analytical Sciences (Moscow, 2006). From left to right: Vice president of the Russian Academy of Sciences academician, N. A. Plate; professor H. Akaiwa (Japan); corresponding member of the Russian Academy of Sciences, B. Ya. Spivakov; academician, Yu. A. Zolotov; and Mayor of Moscow City, Yu. M. Luzhkov. Picture provided by the author

In addition, the *Danube Symposium on Analytical Chemistry*, a conference on the analytical chemistry of the countries of the Black Sea basin, as well as other events, were also held.

10.5 Russian Researchers Abroad

We are talking here about researchers who left Russia to live and work in other countries. Some specialists who studied and worked in Russia are now active analysts in Canada, Australia, Israel, etc. Many of them maintain ties with their homeland, participate in scientific conferences organized in Russia, and sometimes conduct joint research.

For example, I. Kuselman, works in the Israel Physical Laboratory (Fig. 10.3). He is a well-known specialist in the quality assurance of chemical analysis, in metrology. In the same Jerusalem laboratory worked F. Sherman, a former employee of the N. D. Zelinskii Institute of Organic Chemistry, who was engaged in work on aquametry. Another scientist at the same laboratory was Ya. Tur'yan, from Krasnodar. He was a doctor of chemical sciences, well-known in the field of electrochemical analysis methods, and author of a number of books.

Fig. 10.3 Il'ya Isaevich Kuselman, an employee of the Physical Laboratory, Jerusalem, Israel. Picture provided by I. I. Kuselman

Fig. 10.4 Pavel Nikolaevich Nesterenko, University Professor at Hobart, Australia. Picture provided by P. N. Nesterenko

O. Krokhin, former employee of the Division of Analytical Chemistry of the M. V. Lomonosov Moscow State University is known (in Canada) in the field of mass spectrometric analysis of bio-objects. The Division presented Australia with a famous chromatographer, professor of the University of Tasmania, P. Nesterenko (Fig. 10.4), and T. Komarova, who conducts work on passive samplers in Brisbane A. Smirnova is engaged in microfluid analytical systems the University of Tokyo. After leaving for other countries, D. Katskov (atomic absorption, South Africa), Yu. Kazakevich (chromatography and other methods, USA), and others continued their scientific work in the field of chemical analysis.

It is especially necessary to mention Alexander Makarov (Fig. 10.5). After leaving the Moscow Engineering Physics Institute and working for Thermo Scientific in Bremen, he developed the original Orbitrap mass spectrometer (see Chap. 2). Anatolii Verenchikov, a former employee of the Institute of Analytical Instrumentation of the Russian Academy of Sciences, developed a new mass spectrometer based on a multi-path, time-of-flight mass analyzer.

Fig. 10.5 Aleksandr Alekseevich Makarov, the main creator of Orbitrap mass spectrometers. Photo taken from the Internet

10.6 Experience of Organizing Research in Russia

A key unifying, organizational role in Russian analytical chemistry is played by the Scientific Council on Analytical Chemistry of the Russian Academy of Sciences [3]. It was established in 1940 and until 1970 was called the Commission on Analytical Chemistry of the USSR Academy of Sciences. Currently, the council includes about 200 leading Russian analysts who were invited to the council by its chairman, who in turn was appointed by the Academy of Sciences. The council has substantive commissions and several regional branches.

Areas of council activity are:

- Convening scientific conferences (Fig. 10.6).
- Organizing seminars in various cities.
- Participation in the organization of the annual exhibition known as "AnalyticaExpo" (Moscow).
- Publishing activity (preparation of a series of monographs and collections).
- Harmonizing terminology.
- Improving the teaching standards of analytical chemistry and facilitating the retraining of working analysts.
- Improving analytical services.
- Ensuring the quality of analyses.
- Participating in decision about state tasks (expertise, forecasts, etc.).
- Ensuring international activity.
- Organizing competitions and awarding prizes.
- Popularizing analytical chemistry.

Below are the subjects forming the content of scientific conferences which are regularly held by the council: spectroscopic (optical) methods; X-ray methods; chromatography; electroanalytical methods; analytical instruments; and analysis of environmental objects. Earlier, the council also organized regular conferences on organic analytical reagents, liquid–liquid extraction, analysis of organic substances, radioactivation analysis, test methods, and analysis of high-purity substances.

Fig. 10.6 *International Conference on Coordination Chemistry* (Moscow, 1973). Photo provided by the author

We list here a series of monographs, including those collectively published under the aegis of the Scientific Council: *Analytical Chemistry of Elements* (54 volumes, 1960–1990, Nauka Publishing House); *Methods of Analytical Chemistry* (15 volumes, 1970–1980, Khimiya Publishing House); *Analytical Reagents* (10 volumes, 1970–1980, Nauka); and *Problems of Analytical Chemistry* (20 volumes, 1990–2015, Nauka).

Applied, especially industrial, analytical laboratories are united by the Association of Analytical Centers "Analytica." It deals with the issues of laboratory

Fig. 10.7 Chairmen of the Scientific Council of the Russian Academy of Sciences on Analytical Chemistry, academicians A. P. Vinogradov (1941–1954), I. P. Alimarin (1954–1989), and Yu. A. Zolotov (since 1989)

accreditation, standardization of procedures, metrology of chemical analysis, and advanced training of laboratory workers. As already mentioned above, the association is active in the international arena as well (Fig. 10.7).

Until the end of the 1980s branch structures that united the workers of the relevant analytical services were functioning well, e.g., analysts working in the geological industry were cared for by the Scientific Council on Analytical Methods, which operated on the basis of the All-Union Institute of Mineral Raw Materials. The leading scientific research institutes of industries, agriculture, health, nature protection, etc., all held seminars for the workers of analytical laboratories, supplied them with procedures, organized a system of centralized purchasing of instruments, etc.

References

1. Spivakov, B.Ya., Shtykov, S.N.: Zavod. Labor. Diagnos. Mater. **82**, 66 (2016)
2. Baranovskaya, V.B., Boldyrev, I.V., Karpov, Y.A., Fedotov, P.S.: Zavod. Labor. Diagnos. Mater. **83**, 5 (2017)
3. Zolotov, Y.A.: Analytical Chemistry: Science, Applications, People, 149. Nauka, Moscow (2009) (in Russian)

Printed in the United States
By Bookmasters